T0332523

ATMOSPHERIC THERMODYNAMICS
Elementary Physics and Chemistry

This textbook presents a uniquely integrated approach in linking both physics and chemistry to the study of atmospheric thermodynamics.

The book begins by explaining the classical laws of thermodynamics, and discusses Gibbs energy and the elementary kinetic theory of gases with special applications to the atmosphere. Individual chapters focus on various fluid systems, including vapor pressure over flat and curved surfaces of pure liquids and solutions, and examine the vertical dependence of temperature and pressure for environmental sounding and moving air parcels. Recognizing the increasing importance of chemistry in the meteorological and climate sciences, a chapter is devoted to chemical thermodynamics and contains an overview of photochemistry.

Although students are expected to have some background knowledge of calculus, general chemistry and classical physics, the book provides set-aside refresher boxes as useful reminders. It contains over 100 diagrams and graphs to supplement the discussions. It also contains a similar number of worked examples and exercises, with solutions included at the end of the book. It is ideal for a single-semester advanced course on atmospheric thermodynamics, and will prepare students for higher-level synoptic and dynamics courses.

GERALD R. NORTH received a Ph.D. in Physics from the University of Wisconsin in 1966, and has been a Distinguished Professor of Atmospheric Sciences at Texas A&M University for over 20 years. His notable research career includes receiving the Outstanding Publication Award, National Center for Atmospheric Research in 1975, the Exceptional Scientific Achievement Medal for NASA in 1983, and the Jule G. Charney Award from the American Meteorological Society in 2008.

TATIANA L. ERUKHIMOVA received a Ph.D. in Physics from the Institute of Applied Physics, Russian Academy of Sciences, in 1999, and is now a Lecturer in the Department of Physics at Texas A&M University. Her areas of research include large-scale and mesoscale atmospheric transport and mixing, atmospheric wave dynamics, atmospheric ozone, and remote sensing.

ATMOSPHERIC THERMODYNAMICS
Elementary Physics and Chemistry

GERALD R. NORTH AND TATIANA L. ERUKHIMOVA

Texas A & M University, USA

CAMBRIDGE
UNIVERSITY PRESS

CAMBRIDGE
UNIVERSITY PRESS

University Printing House, Cambridge CB2 8BS, United Kingdom

One Liberty Plaza, 20th Floor, New York, NY 10006, USA

477 Williamstown Road, Port Melbourne, VIC 3207, Australia

314-321, 3rd Floor, Plot 3, Splendor Forum, Jasola District Centre, New Delhi-110025, India

79 Anson Road, #06-04/06, Singapore 079906

Cambridge University Press is part of the University of Cambridge.

It furthers the University's mission by disseminating knowledge in the pursuit of
education, learning and research at the highest international levels of excellence.

www.cambridge.org
Information on this title: www.cambridge.org/9780521899635

© G. North and T. Erukhimova 2009

First published 2009

A catalogue record for this publication is available from the British Library

ISBN 978-0-521-89963-5 Hardback

We both wish to dedicate this book to our families, who have kindly endured our long preoccupation with the project.

Contents

Preface

This book is intended as a text for undergraduates in the atmospheric sciences. The students are expected to have some calculus, general chemistry and classical physics background although we provide a number of refreshers for those who might have less experience or need reminders. Our students have also had a survey of the atmospheric sciences in a qualitative course at freshman level. The primary aim of the book is to prepare the student for the synoptic and dynamics courses that follow. We intend that the student gain some understanding of thermodynamics as it applies to the elementary systems of interest in the atmospheric sciences. A major goal is for the students to gain some facility in making straightforward calculations. We have taught the material in a semester course, but in a shorter course some material can be omitted without regrets later in the book. The book ends with two chapters that are independent of one another: Chapter 8 on thermochemistry and Chapter 9 on the thermodynamic equation.

This book is the result of teaching an introductory atmospheric thermodynamics course to sophomores and juniors at Texas A&M University. Several colleagues have taught the course using earlier versions of the notes and we gratefully acknowledge Professors R. L. Panetta, Ping Yang, and Don Collins as well as the students for their many helpful comments. In addition, we have received useful comments on the chemistry chapter from Professors Sarah Brooks, Gunnar Schade, and Renyi Zhang. We also thank Professor Kenneth Bowman for many fruitful discussions. We are grateful for financial support provided by the Harold J. Haynes Endowed Chair in Geosciences.

1

Introductory concepts

The atmosphere is a compressible fluid, and the description of such a form of matter is usually unfamiliar to students who are just completing calculus and classical mechanics as part of a standard university physics course. To complicate matters the atmosphere is composed of not just a single ingredient, but several ingredients, including different (mostly nonreactive) gases and particles in suspension (aerosols). Some of the ingredients change phase (primarily water) and there is an accompanying exchange of energy with the environment. The atmosphere also interacts with its lower boundary which acts as a source (and sink) of friction, thermal energy, water vapor, and various chemical species. Electromagnetic radiation enters and leaves the atmosphere and in so doing it warms and cools layers of air, interacting selectively with different constituents in different wavelength bands.

Meteorology is concerned with describing the present state of the atmosphere (temperature, pressure, winds, humidity, precipitation, cloud cover, etc.) and in predicting the evolution of these primary variables over time intervals of a few days. The broader field of atmospheric science is concerned with additional themes such as climate (statistical summaries of weather), air chemistry (its present, future, and history), atmospheric electricity, atmospheric optics (across all wavelengths), aerosols and cloud physics. Both the present state of the bulk atmosphere and its evolution are determined by Newton's laws of mechanics as they apply to such a compressible fluid. *Dynamics* is concerned primarily with the motion of the atmosphere under the influence of various natural forces. But before one can undertake the study of atmospheric dynamics, one must be able to *describe* the atmosphere in terms of its primary variables. An essential tool needed in this description is thermodynamics, which helps relate the fundamental quantities of pressure, temperature and density as atmospheric parcels move from place to place. Such parcels contract and expand, their temperatures rise and fall; water changes phase, back and forth from vapor to liquid to ice; chemical constituents react,

1

etc. The key to understanding these changes lies in applications of the laws of thermodynamics which relate these changes to fluxes of energy and other less familiar functions which will be introduced as needed.

1.1 Units

The units used in atmospheric science are the *Standard International* (SI) units. These are essentially the MKS units familiar from introductory physics and chemistry courses. The unit of length is the *meter*, abbreviated m; that for mass is the *kilogram*, abbreviated kg; and for time the unit is the *second*, abbreviated s. The units for velocity then are $\mathrm{m\,s^{-1}}$. The unit of force is the *newton* ($1\,\mathrm{kg\,m\,s^{-2}}$, abbreviated N). Tables 1.1–1.5 show the SI units for some basic physical quantities commonly used in atmospheric science.

The unit of pressure, the *pascal* ($1\,\mathrm{N\,m^{-2}} = 1\,\mathrm{Pa}$), is of special importance in meteorology. In particular, atmospheric scientists like the *millibar* (abbreviated mb), but in keeping with SI units more and more meteorologists use the *hectopascal* (abbreviated hPa, $100\,\mathrm{Pa} = 1\,\mathrm{mb}$). The *kilopascal* ($1\,\mathrm{kPa} = 10\,\mathrm{hPa}$) is the formal SI unit and some authors prefer it. *One atmosphere* (abbreviated 1 atm) of pressure is

$$
\begin{aligned}
1\,\mathrm{atm} &= 1.013\,\mathrm{bar} \\
&= 1013.25\,\mathrm{mb} \\
&= 1013.25\,\mathrm{hPa} \\
&= 101.325\,\mathrm{kPa} \\
&= 101325\,\mathrm{Pa} \\
&= 1.01325 \times 10^5\,\mathrm{Pa}
\end{aligned}
\tag{1.1}
$$

and $1\,\mathrm{mb} = 1\,\mathrm{hectopascal} = 100\,\mathrm{Pa}$. In some operational contexts and often in the popular media one still encounters pressure in inches of mercury (in Hg) or millimeters of mercury (mm Hg); $1\,\mathrm{atm} = 760.000\,\mathrm{mm\,Hg} = 29.9213\,\mathrm{in\,Hg}$.

The *dimensions* of a quantity such as density, ρ, can be constructed from the fundamental dimensions of length, mass, time and temperature, denoted by $\mathbf{L, M, T, Temp}$ respectively. The dimensions of density, indicated with square brackets $[\rho]$, are $\mathbf{M\,L^{-3}}$. In the SI system the *units* are $\mathrm{kg\,m^{-3}}$. Many quantities are pure numbers and have no dimension; examples include arguments of functions such as sine or log. The radian is a ratio of lengths and is considered here to be dimensionless.

Temperature in SI units is expressed in degrees Celsius, e.g. $20\,^\circ\mathrm{C}$; or Kelvin, e.g. $285\,\mathrm{K}$. We say "285 kelvins" and omit writing the superscript "o" when

Table 1.1 *Useful numerical values*

Universal	
gravitational constant	$6.673 \times 10^{-11}\,\mathrm{N\,m^2\;kg^{-2}}$
universal gas constant (R^*)	$8.3145\,\mathrm{J\,K^{-1}\,mol^{-1}}$
Avogadro's number (N_A) [gram mole]	6.022×10^{23} molecules $\mathrm{mol^{-1}}$
Boltzmann's constant (k_B)	$1.381 \times 10^{-23}\,\mathrm{J\;K^{-1} molecule^{-1}}$
proton rest mass	$1.673 \times 10^{-27}\,\mathrm{kg}$
electron rest mass	$9.109 \times 10^{-31}\,\mathrm{kg}$
Planck's constant	$6.626 \times 10^{-34}\,\mathrm{J\,s}$
speed of light in vacuum	$3.00 \times 10^8\,\mathrm{m\,s^{-1}}$
Planet Earth	
equatorial radius	$6378\,\mathrm{km}$
polar radius	$6357\,\mathrm{km}$
mass of Earth	$5.983 \times 10^{24}\,\mathrm{kg}$
rotation period (24 h)	$8.640 \times 10^4\;\mathrm{s}$
acceleration of gravity (at about 45°N)	$9.8067\,\mathrm{m\,s^{-2}}$
solar constant	$1370\,\mathrm{W\,m^{-2}}$
Dry air	
gas constant (R_d)	$287.0\,\mathrm{J\,K^{-1}kg^{-1}}$
molecular weight (M_d)	$28.97\,\mathrm{g\,mol^{-1}}$
speed of sound at 0 °C, 1000 hPa	$331.3\,\mathrm{m\,s^{-1}}$
density at 0 °C and 1000 hPa	$1.276\,\mathrm{kg\,m^{-3}}$
specific heat at constant pressure (c_p)	$1004\,\mathrm{J\,K^{-1}\,kg^{-1}}$
specific heat at constant volume (c_v)	$717\,\mathrm{J\,K^{-1}\,kg^{-1}}$
Water substance	
molecular weight (M_w)	$18.015\,\mathrm{g\,mol^{-1}}$
gas constant for water vapor (R_w)	$461.5\,\mathrm{J\,K^{-1}kg^{-1}}$
density of liquid water at 0 °C	$1.000 \times 10^3\,\mathrm{kg\,m^{-3}}$
standard enthalpy of vaporization at 0 °C	$2.500 \times 10^6\,\mathrm{J\,kg^{-1}}$
standard enthalpy of fusion at 0 °C	$332.7\,\mathrm{kJ\,kg^{-1}}$
specific heat of liquid water	$4179\,\mathrm{kJ\,kg^{-1}\,K^{-1}}$
STP	$T = 273.16\,\mathrm{K}, p = 1013.25\,\mathrm{hPa}$

using degrees kelvin. In operational meteorology we sometimes find temperature expressed in degrees Fahrenheit, e.g. 70 °F.

Each side of an equation must have the same dimensions. This principle can often be used to find errors in a problem solution. The argument of functions such as the exponential has to be dimensionless.

1.2 Earth, weight and mass

The Earth is an oblate spheroid, with slightly larger diameter in the equatorial plane than in a meridional (pole-to-pole) plane. The distance from the center to the poles

Table 1.2 *Selected physical quantities and their units*

Quantity	Unit	Abbreviation
mass	kilogram	kg
length	meter	m
time	second	s
force	newton	N
pressure	pascal	$Pa = N\,m^{-2} = 0.01\,hPa$
energy	joule	J
temperature	degree Celsius	°C
temperature	degree Kelvin	K
speed		$m\,s^{-1}$
density		$kg\,m^{-3}$
specific heat		$J\,kg^{-1}\,K^{-1}$

Table 1.3 *Greek prefixes applied to SI units*

Prefix	Numerical meaning	Example	Abbreviation
nano	10^{-9}	nanometer	nm
micro	10^{-6}	micrometer	μm
milli	10^{-3}	millimeter	mm
centi	10^{-2}	centimeter	cm
hecto	10^{2}	hectopascal	hPa
kilo	10^{3}	kilogram	kg
mega	10^{6}	megawatt	MW
giga	10^{9}	gigawatt	GW
tera	10^{12}	terawatt	tW

Table 1.4 *Selected conversions to SI units*

Quantity	Conversion
energy	$4.186\,J = 1\,cal$
	$1\,kWh = 3.6 \times 10^{6}\,J$
pressure	$1\ atm = 760\,mm\,Hg$
	$1\,atm = 29.9213\,in\,Hg$
distance	$1\,m = 3.281\,ft$
temperature	$T(K) = T(°C) + 273.16$
	$T(°F) = \frac{9}{5}T(°C) + 32$
	$T(°C) = \frac{5}{9}(T(°F) - 32)$

Table 1.5 *Some relationships*
between SI units

Quantity	Equivalent
1 N	$1 \, \text{kg m s}^{-2}$
1 J	$1 \, \text{kg m}^2 \, \text{s}^{-2}$
1 Pa	$1 \, \text{N m}^{-2}$

is 6356.91 km and the radius in the equatorial plane is 6378.39 km. About two thirds of the Earth's atmosphere lies below 10 km above the surface, hence the atmosphere and the oceans (depth averaging 4–5 km) only form a very thin skin of about 1/60 the radius of the sphere.

The *weight* of a mass is the force applied to that mass by the force of gravity. It may be expressed as the mass in kilograms times the acceleration due to gravity, $g = 9.81 \, \text{m s}^{-2}$:

$$\boxed{\mathcal{W} = \mathcal{M}g} \quad \text{[weight and mass].} \tag{1.2}$$

Weight is expressed in *newtons*, abbreviated N; $N = \text{kg m s}^{-2}$. The acceleration due to gravity varies slightly with altitude above sea level

$$g_z = g_0(1 - 3.14 \times 10^{-7}z) \ z \text{ in meters.} \tag{1.3}$$

There is also a slight variation ($< 0.3\%$) with latitude due to the ellipsoidal shape of the Earth (due to both centrifugal force and the equatorial bulge). In most meteorological applications these variations are negligible. However, in calculations of satellite orbits such variations are extremely important.

Example 1.1 The density of water in old fashioned units (cgs) is

$$\rho_{\text{water}} = \frac{1 \text{ gram}}{\text{cm}^3}.$$

To express this in SI units, we can multiply by

$$1 = \frac{1 \text{ kg}}{10^3 \text{ gram}} = \text{unity (no dimension).}$$

We obtain:

$$\rho_{\text{water}} = 1\frac{\text{kg}}{10^3 \text{ cm}^3} = 1\frac{\text{kg}}{\text{liter}}.$$

This gives us an intuitive measure of the kilogram. Now we can multiply by

$$1 = \left(\frac{10^2 \text{ cm}}{1 \text{ m}}\right)^3.$$

The final result is

$$\rho_{\text{water}} = 10^3 \text{ kg m}^{-3}.$$

☐

Physics refresher: vertical motion of a particle The acceleration due to gravity is $g = 9.81 \text{ m s}^{-2}$. A particle falling from a height z_0 with no initial velocity, has a velocity $-gt$ after time t. After the same time interval it will have fallen $\frac{1}{2}gt^2$ meters. These are both obtained by simple integration:

$$v_z(t) = \int_0^t -g \, dt = -g \int_0^t dt = -gt, \quad \text{since } g = \text{constant} \tag{1.4}$$

and

$$z(t) - z(0) = \int_0^t v_z(t) \, dt = \int_0^t -gt \, dt = -\frac{1}{2}gt^2. \tag{1.5}$$

More vertical motion mechanics The minimum *work* necessary to lift a particle a vertical distance z against the force of gravity is force × distance = $\mathcal{M}gz$ ($\mathcal{M}g$ is the weight or vertical force necessary to lift the mass without accelerating it). This work done in lifting the particle is equal to the change in its *potential energy* $\mathcal{M}gz$. If the particle is released, work will be done by the gravitational force applied to the particle. The *kinetic energy* of the particle during its fall is $\frac{1}{2}\mathcal{M}v^2$. The conservation of mechanical energy says the sum of these two forms of energy is conserved: $E = \text{PE} + \text{KE} = \text{constant}$, or more explicitly

$$\boxed{\frac{1}{2}\mathcal{M}v_0^2 + \mathcal{M}gz_0 = \frac{1}{2}\mathcal{M}v_t^2 + \mathcal{M}gz_t} \tag{1.6}$$

where the subscript 0 denotes the initial time and the subscript t denotes evaluation at a later time.

The conservation law is derived by first writing Newton's Second Law:

$$\mathcal{M}\frac{dv_z}{dt} = F_z = -\mathcal{M}g. \tag{1.7}$$

Now multiply through by $v \, dt$ and integrate with respect to t. The left-hand side becomes

$$\int_0^t v\mathcal{M}\frac{dv}{dt}\,dt = \frac{1}{2}\mathcal{M}v_t^2 - \frac{1}{2}\mathcal{M}v_0^2. \tag{1.8}$$

On the other side of Newton's equation we have

$$\int_0^t -v\mathcal{M}g\,dt = \int_{z_0}^{z_t} -\mathcal{M}g\,dz = -\mathcal{M}g(z_t - z_0). \tag{1.9}$$

Equating these expressions gives our answer (1.6).

1.3 Systems and equilibrium

Thermodynamics is the study of *macroscopic* or *bulk* systems of masses and their interrelations under conditions of steady state (no dependence on time). By macroscopic we mean the system contains large numbers of individual molecules (within a few orders of magnitude of a *mole*[1] which contains 6.02×10^{23} molecules). We call these states *equilibrium states* if they are not only time independent but also stable under small perturbations. Thermodynamic states are describable by a set of dimensional quantities which we refer to as *coordinates*. Thermodynamics is concerned with the changes in energy-related quantities (certain of the coordinates) when the system undergoes a transition from one state to another. A *thermodynamic system* is a region of space containing matter with certain *internally uniform properties* such as pressure and temperature. We will be concerned with the interior of the system and the variables (coordinates) that characterize it. For example, a mass of pure gas (only one chemical species) contained in a vessel may be characterized by the pressure it exerts on the walls of the vessel, the volume of the vessel and the temperature (p, V, T). These comprise the complete set of *thermodynamic coordinates* for this particular system. For more general situations such as mixtures of species or phases, the coordinates necessary to describe the state have to be determined experimentally. It is important to note that an individual thermodynamic system is uniform in its interior. There are no gradients of pressure or temperature, for example, inside the system.[2]

---

[1] The mole is an SI unit defined as the number of carbon atoms in a mass of 0.012 kg of pure carbon. The number of moles of a substance is the number of multiples of this number (known as Avogadro's number: $N_A = 6.02 \times 10^{23}$). In formulas the unit is designated as "mol."

[2] Note that a column of air in the atmosphere is not a simple thermodynamic system because its pressure and temperature vary with altitude. However, it is convenient to consider the column as composed of thin slabs, each of which contains substance with approximately uniform temperature, pressure and composition. Then each individual slab may be considered as a simple thermodynamic system for many purposes.

1.3.1 Examples of thermodynamic systems

Gas in a vessel Suppose a container holds a gas of uniform chemical composition. Let the walls of the container be thermally insulating and let the volume be fixed. In a very short time after fixing these conditions the gas will come to values of temperature and pressure that are uniform throughout and independent of the shape of the container. This is the simplest thermodynamic system in a state of equilibrium.

A second case is where the container's walls are held at a fixed temperature and the pressure is allowed to vary. Equilibrium will be established such that the temperature of the gas becomes equal to that of the surrounding walls, the volume is given and the pressure comes to some value that we can estimate.

A third case is where the container has a frictionless movable piston that is pushed upon externally by a fixed pressure (such as the atmospheric pressure). This means that the pressure in the vessel is held fixed along with that of the temperature. The piston will shift in such a way to make the pressure inside equal to that outside, and the volume will change until all these conditions are met.

Our gas might not be homogeneous, but instead it might be composed of a mixture of chemically noninteracting gases, such as those in our atmosphere: nitrogen, oxygen and argon. We still have a thermodynamic system as long as the composition does not vary from location to location or from time to time. In each of the above cases let two of the following be fixed: volume, temperature, or pressure. Then the remaining variable is allowed to find its equilibrium value. Note that once in equilibrium, the variables or coordinates are uniform throughout the vessel.

Two-phase system Suppose we have a liquid of uniform chemical composition such as water in our vessel and vacuum above the liquid surface. Let the temperature and volume be fixed. After a sufficient adjustment time some liquid will have evaporated into the volume above its surface and an equilibrium will be established (the flux of water molecules leaving the surface becomes equal to the flux entering and sticking to the surface). There will be a gas pressure exerted on the walls by the vapor that evaporated from the liquid surface. This is a two-phase system with liquid and gaseous phases, but only one *component* (water) which depicts the number of distinct chemical species. The pressure throughout will be uniform (ignore the pressure increase as a function of depth due to gravity in the liquid). The temperature will also be uniform throughout both phases of the system. This two-phase configuration is also a thermodynamic system. The system can be made to pass through changes in volume, temperature, etc., to establish new thermodynamic states of equilibrium. Note that the temperature and pressure are uniform throughout but the density varies from one phase to the other. As we shall see in a later chapter there is another quantity that is also uniform in the two-phase system called the specific Gibbs energy (chemical potential in the chemical literature when expressed as molar Gibbs energy). It acts as an intensive variable (see Section 1.5) in such multicomponent systems similarly to pressure or temperature.

Aqueous solutions Imagine a vessel filled with water (at a fixed temperature and pressure) and some salt is placed in the liquid. If we continue to put more salt into the water eventually some salt will remain in crystal form sinking to the bottom (but ignore gravity otherwise). We will have established an equilibrium between the *saturated* saline solution and the precipitated crystalline salt. A change in temperature will result in a new equilibrium state with a different concentration of salt in solution (concentration of a species in solution is another thermodynamic coordinate). This is an example of a thermodynamic system. Variations on this include allowing the water vapor above the liquid to be in equilibrium with the saline solution. The presence of salt in solution will alter the vapor pressure over the liquid surface (as well as the freezing temperature). As the temperature changes the vapor pressure will change, etc.

Chemical equilibrium Imagine a gaseous mixture in our vessel at fixed temperature and pressure composed of O and O_2. There will be a reaction

$$O + O_2 + M \rightarrow O_3 + M, \tag{1.10}$$

where M is a background molecule used to carry away momentum (e.g., O_2, N_2 or Ar in the atmosphere).[3] Some ozone will decay and after a while there will be an equilibrium established and the reaction can be written:

$$O + O_2 + M \leftrightharpoons O_3 + M. \tag{1.11}$$

The amount of *reactants* (the left-hand side) may be more than the amount of *products* (right-hand side) for a given temperature. But as the temperature is changed the balance may shift. This is a thermodynamic system. The ratio of O_2 to O_3 is now a thermodynamic coordinate along with $T, p, V, \mathcal{M}_{\text{total}}$.

Of course, there are many other types of thermodynamic systems, and we will encounter several of them in due course.

Everything outside the system which may affect the system's behavior is called the *surroundings*. In atmospheric science, we can often approximate an infinitesimal volume of gas embedded in the natural atmosphere as having uniform interior properties. When appropriate, such an infinitesimal volume element can be considered as a thermodynamic system. In many cases the "infinitesimal volume element" might be as big as a classroom or sometimes as small as a cubic centimeter depending on the application.

A thermodynamic system composed of a very large mass is called a *reservoir* and is characterized by a temperature, T_R. If a finite system is brought into contact with the reservoir through a *diathermal* membrane (one which allows the passage

---

[3] Energy and momentum cannot be conserved simultaneously when two bodies go to one with a release of energy. A third body in the collision can provide the means of conserving both.

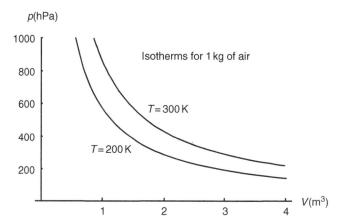

Figure 1.1 Isotherms for 1 kg of dry air taken as an ideal gas. The vertical coordinate is pressure in hPa, the abscissa is volume in m^3. Upper curve, 300 K; lower curve, 200 K.

of thermal energy,[4] but not mass), the smaller system will adjust the values of its coordinates (for a gas, p, V, T) to new values, while the reservoir does not change its state appreciably (this actually defines how massive the reservoir has to be). The system is said to come into thermal equilibrium with the reservoir (its temperature approaches that of the reservoir). In the case of a gaseous system, experiments have shown that there is a locus of pairs of values (V, p) for which the system is in equilibrium with a given reservoir – in other words, a curve $p = p_T(V)$ in the V–p plane. To put it another way, if our system has a certain fixed volume, then when it is brought into contact with the reservoir of temperature T, the pressure will always come to the same value, $p = p_T(V)$. As we do the experiment with different control volumes we can sweep out the locus of points in the V–p plane. This curve is called the *isotherm* of the system for that reservoir temperature (Figure 1.1). The isotherm represents a series of equilibrium states that can occur while the system is in contact with the reservoir (of fixed temperature). For example, the volume might be forced to alter by a change in the wall dimension (e.g., a piston can have different positions in a cylinder which contains the system in question). In this case the pressure will change as a function of volume along the isotherm. While we could invent an algorithm based upon a series of reservoirs of different temperatures to build a temperature scale, it will suffice for our present purposes simply to use the familiar thermometer.

---

[4] Thermal energy refers to the microscopic motion of molecules in the system. When in diathermal contact, the thermal energy of molecules from one system can pass from the system to its neighbor through collisions. In time the thermal energies of the two systems will equalize. More on this in later chapters. The transfer of thermal energy is loosely referred to as *heat transfer*.

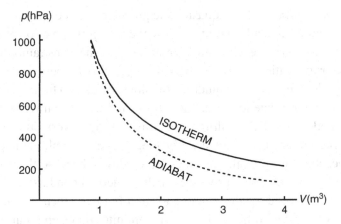

Figure 1.2 Isotherm and adiabat for 1 kg of dry air taken as an ideal gas. The upper curve (solid line) is the 300 K isotherm and the dashed curve is the adiabat passing through the 300 K isotherm at $V = 1\,\mathrm{m}^3$. The vertical coordinate is pressure in hPa, the abscissa is volume in m^3.

A system can also be in equilibrium when isolated (no mass or thermal energy flows into or out of the system) from other systems. We call this an *isolated system*. It can have coordinates just as in the case of a system in contact with a reservoir. We call the locus of values of pressure in the isolated system for different volumes of the system *adiabats* (Figure 1.2). We could find the temperature of the isolated system at fixed values of p and V by bringing it into contact with different reservoirs until we find one which does not cause the coordinates of the system to change. The system has the same temperature as this reservoir. In this way we could map out the locus of points defining the isotherm which crosses the adiabat at the point in question. As a simpler alternative, we could insert a thermometer, whose mass is so small that it will come to equilibrium with the system (which now acts as a reservoir with respect to the tiny thermometer) without disturbing the state of the system appreciably.

States of thermodynamic equilibrium must not involve time. They are steady and only require a knowledge of the thermodynamic coordinates such as temperature, pressure and volume. When the "states" traversed by a system involve the time we cannot use thermodynamic equilibrium states to describe them. Conventional thermodynamics cannot be used to describe what goes on in states that are not in equilibrium.

Certain changes of a system can be made to occur through a sequence of infinitesimally nearby equilibrium states. For example, we might bring the system into contact one at a time with a series of reservoirs of infinitesimally differing temperatures, and at each step we wait for equilibrium to be established. We call this a *quasi-static process*. Such quasi-static processes can be approximated in the laboratory. From a molecular point of view the gas in the interior of the system has to have time during each infinitesimal shift of the constraints to adjust to a new equilibrium with its surroundings. In a gas this is roughly the time for a typical molecule to make a few hundred collisions, but over a finite sized volume it might be more appropriate to use the time for sound waves to traverse the volume several hundred times. This multiple pass traversal time works for pressure, but other properties might take considerably longer. For example, temperature and species concentrations smooth out much more slowly because these differences are smoothed out by diffusive processes such as thermal conduction. Stirring due to turbulence can speed up the homogenization but even then the adjustment is slower than for pressure differences. At each infinitesimal step (waiting for these adjustments) along such a system path, we could reverse direction and retrace the same steps. This is a *reversible process*.

Note that a system may go from one thermodynamic state to another by a path which does not involve such a sequence of thermodynamic states. We call this an *irreversible* change in state. An example of an irreversible process is the case of a system which goes from state A to state B spontaneously, but not from B to A. A concrete example is if two bricks, one hot and one cold, are brought into contact, the result is two warm bricks. This is an irreversible process. Note that it never happens that when we bring two warm bricks into contact we end up with a warm brick and a cold brick (even though energy is conserved).

Reversible processes do not actually occur in nature. So why study them? The reasons are pretty simple. First of all, irreversible processes are nearly impossible to treat theoretically. Secondly, experience has shown that approximating the nearly quasi-static processes that do occur in nature works reasonably well in many cases when we treat them as exactly quasi-static. We proceed then to adopt the philosophy used by practitioners for many years: we will freely approximate many processes in the real atmosphere by idealized reversible analogies in order to obtain numerical results that can be used in practical situations.

1.4 Constraints

An important concept in the study of thermodynamic systems is that of constraints. This notion is best illustrated by example. Consider the gas in a cylinder whose volume is determined by the position of a piston as in Figure 1.3. Several constraints are operative in this case. Most obvious is the position of the piston. It constrains the volume to have a certain value. If the piston is removed by a small amount the constraint is said to be relaxed. Note that a force must be applied

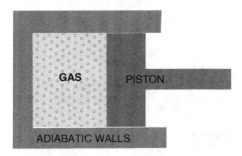

Figure 1.3 Schematic diagram of a gas filled cylinder with adiabatic walls and a movable piston.

(actually relaxed, then gradually reapplied) externally to implement this change in the constraint. If the piston is removed by a small amount, some agent must perform *work* to restore it to its original position. Similarly, the walls that are impervious to the transfer of thermal energy form a constraint. If a leak of thermal energy were to occur, such as on bringing the system into contact with a temperature reservoir at a slightly different temperature, this constraint would be said to have been relaxed and the thermodynamic coordinates of the system will have to be changed to restore the original temperature. Thermodynamic systems are always subject to certain constraints and their nature and number are essential ingredients in the description of the system and its state.

Consider two thermally isolated chambers adjacent to one another separated by a partition. On one side is gas A and on the other is gas B. Let the chambers have the same temperature and pressure. The partition forms a constraint restricting the two gases from mixing. If the partition is removed, the constraint is relaxed and the two systems will pass through nonequilibrium states to their final well-mixed equilibrium state. The irreversible process following removal of the constraint represents one which for ideal gases involves no changes in pressure or temperature, but external work must be performed to restore the original conditions.

1.5 Intensive and extensive quantities

Consider a thermodynamic system. The interior properties of the system are uniform. Now, imagine subdividing the system into two equal parts (say, two warm bricks in contact). If a variable is the same for the two individual parts (e.g., pressure, temperature, chemical composition, density, etc.), the variable is an *intensive variable*. On the other hand, if the thermodynamic variable for each subsystem is proportional to the mass of the constituents in that subsystem (e.g., volume, mass), we call it an *extensive variable*.

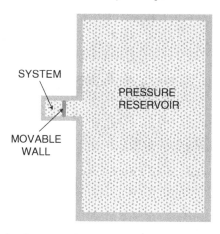

Figure 1.4 Schematic diagram of a gaseous pressure reservoir in contact with a small system. The membrane between the system and the pressure reservoir is movable so that the two systems can adjust their volumes in such a way that the pressures equalize.

Example 1.2 An example of an isolated system is a 1 kg mass of gaseous O_2, confined in a box with thermally insulating walls. Suppose the volume is 1 m^3. This means the density of the gas is 1 kg m^{-3}. If the temperature of the gas is given, say 300 K, then the pressure will be determined (this is an experimental fact). The thermodynamic coordinates of this (pure) system are: V, the volume; $\mathcal{M}$, the mass; T, the temperature; and p, the pressure.   □

Example 1.3 A thermodynamic system might be in thermal equilibrium with a reservoir. In the case of the mass of O_2 gas in a fixed volume of 1 m^3, take the gas to be in thermal contact with a reservoir at 350 K. The pressure will be quite different from the last example.   □

Example 1.4 We might have a pressure reservoir. Consider the box of gas to be in contact with a reservoir with a slidable interface, such that the pressures can equalize between the two systems. Let the system otherwise be insulated thermally from the reservoir and the rest of the universe. If the gas has a given temperature initially, it will expand or contract until its pressure equals that of the reservoir (please let it happen gradually). The volume and temperature of the gas may change in order to establish equilibrium with the pressure reservoir (see Figure 1.4).   □

1.6 System boundaries

Before setting up a problem in thermodynamics it is extremely important to choose the part of the universe you want to call your system. It might be a mass of matter or it might be a certain volume in space. As in the atmospheric examples the mass or volume might be in motion. If we are considering a mass in space with no additional matter allowed to enter or leave this fixed mass we say it is a *closed system*. In the fixed volume case mass might enter or leave. We call this an open system.

Calculus refresher: the exponential function The function $y(x)$ whose derivative is itself is called the exponential function:

$$\frac{dy}{dx} = y. \tag{1.12}$$

Suppose we try $y = a^x$. Then

$$\frac{\Delta y}{\Delta x} = \frac{a^{x+\Delta x} - a^x}{\Delta x} = a^x \frac{a^{\Delta x} - 1}{\Delta x}. \tag{1.13}$$

The factor on the right must tend to unity as $\Delta x \to 0$. It will be more easily seen if we let $\Delta x = 1/N$ where N is an integer. A little rearrangement yields

$$a_\infty = \lim_{N \to \infty} \left(1 + \frac{1}{N}\right)^N \tag{1.14}$$

and the number a_∞ is given the symbol e whose numerical value turns out to be 2.718281. ... To see how the limit comes about call the approximate value of $a_\infty = e_N$. Simple computation gives, $e_5 = 2.48832, e_{10} = 2.59374, e_{100} = 2.70481, e_{1000} = 2.71692,$ and $e_{10000} = 2.71815$. ...

Note that $e^0 = 1$, $e^{-1} = 0.367879\ldots$, and e^x is called the exponential function. We can easily derive a few properties of $y = e^x$. From its definition, $de^x/dx = e^x$, and we can use the chain rule to show that $de^{\alpha x}/dx = \alpha e^{\alpha x}$.

The function $e^{-\alpha x}$ decreases *exponentially* from a value of unity at $x = 0$ to a value

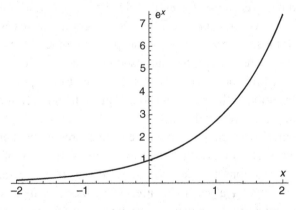

Figure 1.5 The exponential function e^x as a function of x.

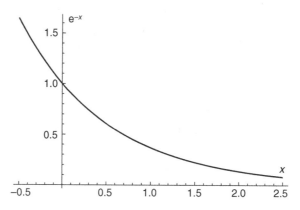

Figure 1.6 The decaying exponential function e^{-x} as a function of x.

of $e^{-1} = 0.367879\ldots$ at $x = 1/\alpha$, which is called the e-*folding distance* if x is a distance or the e-*folding time* or *time scale* if x is a time, see Figures 1.5 and 1.6.

1.7 Thermodynamics and atmospheric science

The plan of this book is to present the subject of thermodynamics in such a way as to help us better understand the atmosphere, but also to give us some rules and methods that can be used in the practical application of atmospheric science. Thermodynamics is a huge subject more than a century old, treated by excellent textbooks in physics, physical chemistry, chemical engineering, mechanical engineering, etc. We cannot possibly cover all the material in these fields. We cannot even cover all the basic theory of thermodynamics in a short course intended for students majoring in atmospheric sciences – especially at the sophomore/junior year level for undergraduates, where not much science and mathematics can be required prerequisites. Some compromises will have to be made. This means those who want to delve deeper into some of the derivations will have to check elsewhere among the many sources listed at the ends of chapters. Sometimes we will limit derivations or justifications to the point that it is clear that enough information is there to determine that such and such a formula can be derived by the methods already discussed.

So what problems in atmospheric science can be addressed by thermodynamics? After all we have seen already that thermodynamics consists of a set of laws applicable under conditions that are so idealized that they are rarely attainable even in the laboratory let alone in nature. The processes that occur in nature are spontaneous and virtually never do we find a system (perhaps our leading application consisting of a parcel of air) in true thermodynamic equilibrium. The answer to these questions is that thermodynamics affords us a useful framework in which some approximate calculations can be conducted. Over the years these approximate results have proven to be useful.

The following is a list of applications that we will work on in this book.

Gaseous mixtures These occur in nature since the atmosphere is composed mainly of several relatively inert gases (at typical atmospheric temperatures anyway) and several other gases which occur in trace amounts (i.e., ozone, methane, carbon dioxide, even water vapor). We will be concerned with how the pressures of these individual gases add up to the total pressure and how the density of a local blob (or parcel) of the gas might differ from that of its surroundings, which might give rise to parcels lifting themselves from their present altitudes to somewhere above, just how high and how fast depending on the parcel's thermodynamic properties and those of the surroundings.

Liquid–vapor equilibria Water exists in all three phases in the atmosphere. We would like to know how the pressure of water vapor above liquid surfaces varies with temperature. We would like to have useful ways of describing the amount of water vapor in the air (humidity) and how this affects the air's buoyancy and how condensation leads to release of thermal energy and therefore changes in buoyancy. Other related issues treatable in thermodynamics include how liquid water droplets grow in humid environments. Does the presence of air affect the vapor pressure above a liquid surface? Does the air dissolve appreciably in the liquid? Does the presence of salt dissolved in the liquid affect the vapor pressure? Does the size of a droplet affect its vapor pressure?

Dynamics of air parcels How can we tell whether a given environmental temperature profile (function of altitude) leads to stable conditions or unstable ones (does the air start to turn over spontaneously)? What is the role of moisture and cloud formation in this process? As parcels rise, they expand and their temperature drops (why?). Does this mean they are denser and they might return? What are the conditions for continued rising? What temperature profiles are likely to lead to severe weather?

Atmospheric chemistry Most chemical reactions in the atmosphere are between trace gases such as ozone and so-called air pollutants, but many occur between natural constituents. What are the criteria for a reaction to proceed one way or another? How are chemical equilibria between reactants and products established and how do these equilibrium concentrations vary as the temperature varies?

Notes

A complete bibliography is given at the end of the book. All university level physics books contain a few chapters on thermodynamics; the numerous editions of Sears and Zemanski as well as those of Halliday and Resnick and the one by Giancoli are good examples. Many general chemistry books also contain a good description

of the subject, for example, Whitten, Davis and Peck (1996). The little paperback
titled simply *Thermodynamics* by Enrico Fermi has some marvelous descriptions
of thermodynamic states, equilibria, etc.

An older advanced book which contains specific applications to meteorology
and especially applications to cloud physics is that of Irebarne and Godson (1981).
Two newer books on applications of thermodynamics to atmospheric science and
oceanography are those by Bohren and Albrecht (1998) and Curry and Webster
(1999). Both are pitched at a higher level than the present text and both delve more
deeply into many aspects of the subject.

Notation and abbreviations for Chapter 1

atm	pressure unit, one atmosphere
g, g_z, g_0	acceleration due to gravity, its value as a function of altitude, its value at the surface (m s^{-2})
G	gravitational constant (Table 1.1)
h	height
k_B	Boltzmann's constant (Table 1.1)
L	dimension length
mb,	pressure unit, one millibar (1 mb = 1 hPa)
M	dimension mass
$\mathcal{M}$	mass of a macroscopic object or system (kg)
$\mathcal{M}_E$	mass of the Earth (kg)
N_A	Avogadro's number (Table 1.1)
p	pressure (Pa, hPa)
Pa	unit of pressure, $1\,\mathrm{Pa} = 1\,\mathrm{N\,m}^{-2}$
R, R_G, R_d, R_w, R^*	gas constants: no subscript indicates for an unnamed gas, sometimes explicitly for a specific gas, G; the subscript d for dry air, w for water vapor, and the superscript * for the universal gas constant (Table 1.1)
SI	Standard International system of units
T	dimension time, also period of a repeating process, and temperature
Temp	dimension temperature
v	speed (m s^{-1}), sometimes with a subscript indicating velocity component along a coordinate axis (e.g., v_x)
V	volume of a system (m^3)

Problems

1.1 A useful mathematical model of the vertical dependence of pressure is

$$p(z) = p_0 e^{-z/H}$$

where H is called the *scale height*. H is usually between 8 and 9 km. What is this in thousands of feet? Compare to the altitude of typical jet air flights.

1.2 What is the ratio of the atmospheric scale height (Problem 1.1) to the Earth's radius?

1.3 If p_0 is 1000 hPa, what is a typical pressure in Denver (mile high city)? Use the expression in Problem 1.1 with $H = 8$ km with 1 mile $= 1.6$ km.

1.4 (a) Compute the circumference of the Earth at the Equator in km. (b) How many km are there per degree of longitude at the Equator? (c) How many km are there per degree of longitude at 30°N?

1.5 A numerical model of the atmosphere has horizontal resolution (grid boxes) 2° × 2°. What is the area of one of these boxes at the Equator, and at 30°N?

1.6 Use the conservation of energy to show that for a particle falling under gravity from rest at a height z_0, at height z its velocity is given by

$$v^2 = 2g(z_0 - z)$$

where g is the acceleration due to gravity (9.81 m s^{-2}).

1.7 Newton's Law of Gravity says that the force on a particle of mass $\mathcal{M}$ is

$$F = -\frac{G\mathcal{M}_E\mathcal{M}}{r^2}.$$

Use the fact that the acceleration of gravity at the surface is $g = 9.81$m s^{-2} along with the mass of the Earth $\mathcal{M}_E = 5.983 \times 10^{24}$ kg and the radius of the Earth of 6365 km to compute the gravitational constant G (see Table 1.1).

1.8 A particle falls from outer space to the Earth's surface. Far away its potential energy is zero. At the Earth's surface the potential energy is $-G\mathcal{M}_E\mathcal{M}/R_E = -\mathcal{M}gR_E$ (relative to $r = \infty$) where $\mathcal{M}_E$ is the Earth's mass, R_E its radius and $\mathcal{M}$ is the mass in question (see Table 1.1). What is the velocity of the mass when it strikes the Earth's surface? (This is exactly the vertical velocity it would have to have if it were to escape the Earth's gravity field after being projected upwards from the surface. It is called the *escape velocity*.)

1.9 Suppose a particle is dropped from a height h and it bounces elastically (i.e., no kinetic energy is lost in the collision: $v_{after} = -v_{before}$). How long does the round trip take?

2

Gases

2.1 Ideal gas basics

Gases are a form of matter in which the individual molecules are free to move independently of one another except for occasional collisions. Most of the time the individual molecules are in free flight out of the range of influence of their neighbors. Gases differ from liquids and solids in that the force between neighbors (on the average over time) is very weak, since the intermolecular force is of short range compared to the typical intermolecular distances for the individual gas molecules.

If an imaginary plate is held vertically in a gas as shown in Figure 2.1, there will be a force exerted on the thin plate from each side. The forces on opposite sides of the inserted plane are equal; otherwise, if forces on the opposing sides were unbalanced, the plate would experience an acceleration. The force on the left side of the plate is caused by the reflection of molecules as they hit the left face of the plate and rebound. These impulsive forces are so frequent that the resulting macroscale force is effectively steady. The force is perpendicular to the face and has the same value no matter how the face is oriented. This can be seen by considering the collisions with the wall and the tendency for no momentum to be transferred parallel to the plane surface. The perpendicular component of the force per unit area on the plane is called the *pressure*. Tangential components of the force cancel out (when averaged over many collisions with the wall) and therefore vanish when averages are taken over a large number of collisions with the surface. The pressure has units of newtons per meter squared or $N\,m^{-2}$; 1 $N\,m^{-2}$ is called a pascal, abbreviated Pa. Atmospheric scientists use hPa (hectopascals 1 hPa $= 100\,Pa$) or the equivalent mb (millibars). The more appropriate unit would be the kPa, but this is not used much in practice. A newton is the force necessary to maintain an acceleration of 1 $m\,s^{-2}$ on a 1 kg mass. The units of force may be decomposed to $kg\,m\,s^{-2}$.

The state of a gas is characterized by three quantities: its pressure, p; (mass) density, ρ (note that knowledge of density is equivalent to knowledge of the

FORCES ON A VERTICAL PLATE
IN A GAS

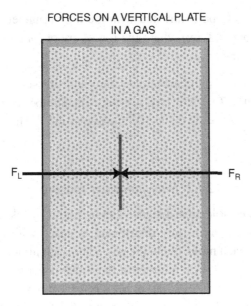

Figure 2.1 Illustration of the forces exerted by a gas on a thin vertically oriented plate. Since the plate does not feel a net force, $\mathbf{F_L} = -\mathbf{F_R}$.

volume for a single-phase system of a given mass, $\rho = \mathcal{M}/V$); and temperature, T (in kelvins). In general, there is a mathematical relationship between the three variables, or thermodynamic coordinates, called the *equation of state*. It is important to remember that the density is uniform throughout the volume of a gas in thermodynamic equilibrium.

The *number density*, n_0, is the number of molecules per unit volume ($[n_0] = $ molecules m^{-3}). The equation of state of an *ideal gas* is given by

$$\boxed{p = n_0 k_B T} \qquad \text{[Ideal Gas Law]} \qquad (2.1)$$

where k_B is called *Boltzmann's constant*:

$$k_B = 1.381 \times 10^{-23} \, \mathrm{J\,K^{-1}\,molecule^{-1}}. \qquad (2.2)$$

Boltzmann's constant is a *universal constant* independent of the molecular species. Almost all gases behave as ideal gases if they are sufficiently dilute. The condition for this is that the molecules spend a large fraction of their time apart from one another so that the intermolecular forces are acting only a small fraction of the time for a given molecule. This will become clearer in the next few sections where some estimates of intermolecular spacings and distances between collisions are compared to the sizes of the molecules. Typical intermolecular forces for neutral molecules are appreciable only over distances of the order of the radius of a molecule. This is to be contrasted with the long-range electrical forces of an ionic species (Coulomb's Law) where the forces are of long range, varying as the inverse square of distance.

Example 2.1 A gas is at standard temperature and pressure (referred to as STP: $p = 1$ atm $= 1013$ hPa, $T = 273.16$ K). What is its number density?

Answer: $n_0 = p/(k_B T) = 2.69 \times 10^{25}$ molecules m^{-3}. This value is called the *Loschmidt number*.   □

2.1.1 Intermolecular spacing

The approximate intermolecular distance can be found by taking the molecules at an instant of time to be uniformly distributed in space with number density n_0. Place a cube around each molecule in the gas. Then each molecule sits at the center of a cube of side length d. The number of these cubes per unit volume is n_0. The volume of one of them is $d^3 = 1/n_0$; or $d \approx 1/n_0^{1/3} = 3.34 \times 10^{-9}$ m $= 3.34$ nm (at STP). Note that the radius of a molecule r_0 is only a few times 10^{-10} m $= 0.1$ nm (several tens of times less than the intermolecular distance). In a liquid or a solid intermolecular distances are on the order of the molecular sizes (see Table 2.1).

Atomic refresher The Bohr atom has radius $a = h^2 \epsilon_0 / \pi m_e Q_e^2$, where $\epsilon_0 = 8.85 \times 10^{-12}$ F m^{-1} (permittivity constant), $h = 6.63 \times 10^{-34}$ J s (Planck's constant), $Q_e = 1.60 \times 10^{-19}$ C (electron charge), $m_e = 9.11 \times 10^{-31}$ kg (electron mass). The Bohr radius for a hydrogen atom is $a = 5.29 \times 10^{-11}$ m $= 0.0529$ nm. Most high school or college chemistry books describe the Bohr model of the hydrogen atom.

2.1.2 Mean free path

The average distance a molecule travels in the gas before collision is called the *mean free path*. To obtain an estimate of the mean free path imagine the background gas particles to be stationary. Take our test molecule of radius r_0 to be moving through the lattice of fixed points used in the last subsection. A collision between our prototype molecule and a background molecule will occur when their centers are within $2r_0$ of each other (Figure 2.2). We can think of the test molecule having radius $2r_0$ and the lattice composed of stationary points (see Figure 2.3). Hence, as the test molecule moves through the lattice it sweeps out a cylinder of radius $2r_0$. In time Δt the volume swept out is $(2r_0)^2 \pi \times v \Delta t$ (number per unit volume times volume). The cross-sectional area of the sweeper is sometimes given the symbol σ_c and called the *collision cross-section* (see Table 2.2). The number of background

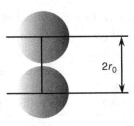

Figure 2.2 At the moment of collision two spherical molecules have a distance between their centers of $2r_0$.

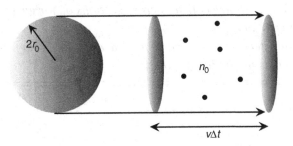

Figure 2.3 As a molecule moves a distance $v\Delta t$, it encounters all the point centers in the volume of the cylinder whose length is $v\Delta t$ and whose cross-sectional area is $4\pi r_0^2$. The number of point centers in the volume is $n_0 4\pi r_0^2 v\Delta t$.

molecule centers in this cylinder is $n_0 \times \sigma_c v\Delta t$ (number per unit volume × volume of the cylinder). Then the number of collisions per unit time (*the collision frequency*) is

$$\boxed{f_{\text{coll}} = n_0\sigma_c v} \quad \text{[collision frequency].} \tag{2.3}$$

We may calculate the average distance between collisions to be the [distance per unit time] = [velocity] × [the time between collisions] (this last factor is the inverse of collision frequency):

$$\lambda = v \times \frac{1}{n_0\sigma_c v} = \frac{1}{n_0\sigma_c} \quad \text{[mean free path, approximate form].} \tag{2.4}$$

The Greek letter λ is used in most texts to denote the mean free path. Actually, it is possible to solve the problem when all the particles are in motion, and the derivation can be found in books on the kinetic theory of gases. The same formula occurs for the mean free path except for an additional factor of $\sqrt{2} = 1.414\ldots$ in the denominator:

$$\boxed{\lambda = \frac{1}{\sqrt{2}n_0\sigma_c}} \quad \text{[mean free path, more exact form].} \tag{2.5}$$

Table 2.1 *Some characteristic lengths in the
kinetic theory of gases at STP (all in units of
nanometers, 10^{-9} m)*

Bohr radius	Intermolecular spacing	Mean free path	Wavelength of yellow light
0.053	3.3	52	500

Table 2.2 *Some collision cross-sections σ_c for gases of
interest in nm^2*
*For interspecies estimates use the average. Data from
Atkins and de Paula (2002).*

Ar–Ar	N$_2$–N$_2$	O$_2$–O$_2$	H$_2$–H$_2$
0.36	0.43	0.40	0.27

For a typical gas $r_0 = 2 \times 10^{-10}$ m (0.2 nm) and thus, $\sigma_c = 4\pi r_0^2 \approx 5 \times 10^{-19}$ m^2 (0.5 nm^2). Using n_0 from above we have $\lambda \approx 5.23 \times 10^{-8}$ m (52.3 nm), which is about an order of magnitude more than the intermolecular spacing at STP (Table 2.1).

Example 2.2 Take as a model for the vertical dependence of number density:

$$n_0(z) = n_0(0)e^{-z/H}$$

where z is the altitude above sea level and H is called the *scale height* (typically 8 km in midlatitudes, but up to 12 km in the tropics). Our exponential model was just cooked up, but it turns out to be a very good approximation. What are typical mean free paths at $z = 0, H, 2H$, taking STP at $z = 0$?
Answer:

$$\lambda = \frac{1}{n(z)\sigma_c} = \frac{e^{z/H}}{n(0)\sigma_c} = \lambda_0 e^{z/H}.$$

Then: $\lambda_{z=0} = 52$ nm, $\lambda_{z=H} = 141$ nm, $\lambda_{z=2H} = 384$ nm. ☐

From the previous example we see that the mean free path is still very small compared to our familiar everyday sizes of things, especially weather phenomena, even at altitudes of several scale heights (well into the stratosphere). When the dimensions of the body of gas (say, a storm or a cold air mass) are large compared to the mean free path, we can ignore the molecular motions and treat the gas as a fluid, obeying the macroscopic laws of fluid mechanics. This is usually the case in atmospheric science up to altitudes of about 100 km.

The mean free path also gives us an idea of the length scales over which transport of properties occurs. If a property is transported via collisions between molecules it has to happen over length scales comparable to the mean free path. Such processes are said to be diffusive, and they are rather slow compared to some fluid motions such as convection.

2.1.3 Pressure from kinetic theory

An intuitive feeling for the pressure of an ideal gas can be gained by considering a gas enclosed in a cubical box of edge length L (see Figure 2.4). Imagine a molecule going left and right across the box and bouncing back at the walls. A round trip takes a time $2L/v_0$ where v_0 is the speed of the molecule. At a reflection it experiences a change of momentum $\Delta(m_0 v_0) = 2m_0 v_0$. Each reflection imposes an impulsive force to the wall (see the physics refresher below). The frequency of such reflections (by the entire box of molecules) is so large that the force is effectively steady (we shall see later that typical molecular speeds are hundreds of $\mathrm{m\,s}^{-1}$). The rate of such impulses by an individual molecule is the change in momentum divided by the time interval between reflections, $2m_0 v_0/(2L/v_0) = m_0 v_0^2/L$. If we suppose that one third of the molecules are going left-right (the others are going up-down and in-out), then the number going left-right is $n_0 L^3/3$. The total force on the wall is

$$F = \left(m_0 v_0^2/L\right)\left(n_0 L^3/3\right). \tag{2.6}$$

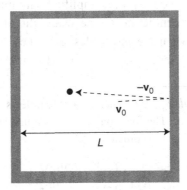

Figure 2.4 A molecule moving left to right with velocity v_0 is elastically reflected by the wall. After the collision the velocity is $-v_0$. The side dimension of the cubical box is L.

This total force on the wall should also be the pressure × the area of the wall which can be expressed as

$$F = \text{pressure} \times \text{area} = pL^2. \tag{2.7}$$

Equating these we have

$$\boxed{p = \frac{1}{3}n_0 m_0 \overline{v_0^2}} \quad [\text{pressure related to } \overline{v_0^2}] \tag{2.8}$$

where the overbar indicates statistical averaging over the probability distribution of molecular speeds, v_0. Since we know the equation of state for an ideal gas (from experiments), $p = n_0 k_B T$, we can use this along with the relationship just found to identify the corresponding coefficients and arrive at the new relationship:

$$\boxed{\frac{1}{2}m_0 \overline{v_0^2} = \frac{3}{2}k_B T} \quad [\text{kinetic energy and Kelvin temperature}]. \tag{2.9}$$

Note that the left-hand side is just the average kinetic energy ($\frac{1}{2}m_0\overline{v_0^2}$) of a molecule in the box. This relationship says that the *temperature* expressed in kelvins is proportional to the average kinetic energy of individual molecules. The coefficient of proportionality is thus determined to be $\frac{3}{2}$ times the Boltzmann constant. Note that when the absolute or Kelvin temperature T is zero, the molecules are at rest ($\overline{v_0^2} = 0$). All thermal motion is presumed zero at this temperature at least in the ideal gas as we have defined it. Actually, there is no such thing as an ideal gas at $0\,K$ (any real gas would have been liquified or solidified well above $0\,K$). Moreover, quantum mechanics tells us that there is motion even at this lowest of low temperatures. Fortunately, we never encounter these low temperatures in meteorology and the Ideal Gas Law virtually always applies to the gases of interest.

While the derivation above is highly simplified with many details omitted, such as flights that are not perpendicular to the walls, distributions of the molecular velocities, collisions between the molecules, etc., these details cancel out in the more rigorous derivation. Hence, the formula and its interpretation are correct.

Physics refresher: impulse force When a particle reflects from a rigid surface it exerts a force on the surface. The force on the wall is found by integrating Newton's Second Law over the time of the collision:

$$\overline{F}_\tau = \frac{1}{\tau} \int_{t-\tau/2}^{t+\tau/2} m_0 \frac{dv(t)}{dt} \, dt$$

where τ is the (very short) time interval during which the molecule is in contact with the wall. Unfortunately, we do not know the value of τ in general. However, over longer periods outside the interval $(t - \tau/2, t + \tau/2)$ the integrand vanishes so the integration period can be extended from $t - T_{rt}/2$ to $t + T_{rt}/2$ where T_{rt} is the length of time for the molecule to make its round trip between collisions with the wall. The force exerted on the wall during a round trip of a single molecule is then

$$\overline{F}_{rt} = \frac{1}{T_{rt}} \int_{t-T_{rt}/2}^{t+T_{rt}/2} m_0 \frac{dv(t)}{dt} dt = \frac{2m_0 v_0}{T_{rt}}.$$

If the round trip time $(T_{rt} = 2L/v_0)$ is short enough (compared to the sluggish wall's response time) the wall will feel a nearly steady force of this magnitude. Now if we add the collisions of all the molecules, as in the derivation of pressure, we can be assured of a steady force perpendicular to the wall.

There are a few loose ends that must be addressed. First, not all molecules are traveling strictly in the x, y and z directions. This can be disposed of by noting that for the wall perpendicular to the x direction only the x component of the motion matters. The y and z components do not affect this wall. Do the collisions one by one cancel their y and z components before and after the collision with the wall? After all, the wall is not a smooth surface at the molecular level. The answer is that over the long run and averaging over many particles this cancellation is complete. Lastly, the molecules do not travel uninterrupted from one end of the box to the other. They go only one mean free path (a few tens of nanometers at STP) before they suffer a collision with another molecule. The solution to this problem lies in the conservation of momentum. After a collision the x component of momentum is conserved for the colliding pair and it is the momentum change at the wall that matters, whether it is the same molecule or not.

In specific applications such as meteorology it is useful to cast the Ideal Gas Law into yet another form by multiplying and dividing by the mass of an individual molecule, m_0:

$$p = m_0 n_0 \left(\frac{k_B}{m_0} \right) T = \rho R T \tag{2.10}$$

where ρ is the mass density $(\rho = m_0 n_0 = \mathcal{M}/V)$ in $\mathrm{kg\,m^{-3}}$, and the gas constant defined by R is k_B/m_0 (note: this is *not* the universal gas constant which is to be defined later). For dry air

$$R_d = 287\,\mathrm{J\,K^{-1}\,kg^{-1}}. \tag{2.11}$$

Note that this definition of the gas constant depends on the mass of individual molecules m_0. Dry air is a mixture of different ideal gases. The value of R_d takes this mixture into account as will be explained shortly.

Table 2.3 *The composition of dry air*

Constituent	Percentage by volume (number count)	Percentage by mass	Molecular weight
nitrogen	78.09	75.51	28.02
oxygen	20.95	23.14	32.00
argon	0.93	1.30	39.94
CO_2	~ 0.03	~ 0.04	44.01

Table 2.3 gives the composition of dry air by volume percentage – this is the ratio of the number density of a substance n_0 to the total number density. The same table also gives the percentage by mass – the ratio ($\times 100$) of the mass of the constituent in a sample to the mass of the whole sample.

Example 2.3 What is the density of a parcel of dry air at 270 K at the 500 hPa level?
Answer: Use $\rho = p/(R_d T)$:

$$\rho = \frac{50\,000\,\text{Pa}}{(287\,\text{J kg}^{-1}\,\text{K}^{-1})(270\,\text{K})} = 0.645\,\text{kg m}^{-3}. \tag{2.12}$$

□

Example 2.4 What is the *root mean square* (rms) speed of a molecule of air at STP?
Answer: We can write

$$\overline{v^2} = 3\frac{k_B}{m_0}T = 3R_d\,T \tag{2.13}$$

$$v_{\text{rms}} = \sqrt{3R_d T} \approx 485\,\text{m s}^{-1}. \tag{2.14}$$

□

Example 2.5 What is the collision frequency of "air" at STP? (*air* is in quotes because we imagine the air composed of a single species whose molecular weight is 29.0, i.e., 29 times the mass of a proton).
Answer: We use the collision frequency formula: $\approx n_0 \sigma_c v_{\text{rms}} = 2.69 \times 10^{25}$ molecules m^{-3} $\times 5 \times 10^{-19}$ m^2 $\times 485$ m s^{-1} $= 6.52 \times 10^9$ collisions s^{-1}. □

Example 2.6 What is the typical number of collisions with a wall perpendicular to the x-axis per unit area per unit time? Let the conditions be STP.
Answer: The number of molecules impinging on the wall from the left is $(n_0/4)(A\bar{v})$ where n_0 is the number density, A is the area of the wall (1 m^2), and $\bar{v}$ is the average speed of the molecules as shown in Section 2.3. $n_0 = 2.69 \times 10^{25}$ molecules m^{-3}

and $\bar{v}$ can be taken to be roughly the v_{rms} of the previous example. Hence, there are about 10^{27} collisions per second with the $1\,\text{m}^2$ wall at STP. So where did the divisor 6 come from? It comes from a careful integration over all the angles, etc. A simple minded (but correct) way of obtaining it is to note that $\frac{1}{3}$ of the molecules are going in the x direction, but only half of these are going in the $+x$ direction. With or without the 6, this is a very large collision rate. $\quad\square$

Math refresher: probability density function (pdf) The quantity $\mathcal{P}(u)\,du$ is the probability of finding the variable u to lie in the range $(u, u + du)$. The probability of finding it to have *any* real value is unity; thus

$$\int_{-\infty}^{\infty} \mathcal{P}(u)\,du = 1.$$

The *mean* value of u is defined to be

$$\mu_u = \int_{-\infty}^{\infty} u\,\mathcal{P}(u)\,du$$

and its *variance* or *mean squared value* is given by

$$\sigma_u^2 = \int_{-\infty}^{\infty} (u - \mu_u)^2 \mathcal{P}(u)\,du.$$

The variable u is called a *random variable*. In treating random variables we consider independent *realizations* of the variable (like drawing values from a hat).

2.2 Distribution of velocities

Obviously the molecules in a box are not moving parallel to the x, y and z directions exclusively. Instead molecules will have instantaneous velocity components v_x, v_y, and v_z. Consider the v_x component for an individual molecule at a given time. The value of v_x will take on a range of values from one time to the next because of collisions with other molecules (it can be thought of as a random variable). Computer simulations suggest that after only a hundred or so collisions per molecule the probability density of values of v_x settles to a steady functional form. After this *equilibration* or *thermalization time* v_x is found to be distributed as:

$$P(v_x) = \frac{1}{\sqrt{2\pi}\,\sigma_{v_x}} e^{-v_x^2/2\sigma_{v_x}^2} \quad \text{[normal distribution]} \qquad (2.15)$$

Gases

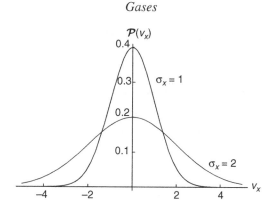

Figure 2.5 Graph of the normal distribution for $\sigma_x = 1$ and for $\sigma_x = 2$, where $\sigma_x = \sqrt{R^*T/M}$. The larger value of σ_x (higher temperature or lower molecular weight) leads to a broader distribution, but still has the same area under it. In the case of the x component of velocity, this means the velocity is expressed in units of $280\,\mathrm{m\,s^{-1}}$ at 300 K. Note that this differs from the rms speed ($485\,\mathrm{m\,s^{-1}}$), since we are only considering one of the three components of velocity.

which is called the normal distribution. The normal (sometimes called the Gaussian) distribution occurs often in nature. It generally comes about when the variable is subjected to a long history of random jolts that add up to its current value. After a long time (many additive increments to the value of the variable) its probability distribution approaches the normal distribution. This can be proved under rather general conditions in mathematical statistics under the heading of the *Central Limit Theorem*. The normal distribution has the familiar bell shape shown in Figure 2.5. This *probability density function* (pdf) is normalized such that

$$\int_{-\infty}^{\infty} P(v_x)\, dv_x = 1 \quad \text{[normalization]}. \tag{2.16}$$

The area under a portion of the curve between two values v_{x_1} and v_{x_2} is the probability of a given molecule having its x component of velocity lying in that range. Obviously the probability of its lying in the range $(-\infty, \infty)$ is unity. The *most probable velocity* is the one for which the pdf is maximum – it is called the *mode* of the distribution; the mode is the value of v_x for which $dP/dv_x = 0$. The average value of v_x is given by

$$\overline{v}_x = \int_{-\infty}^{\infty} v_x P(v_x)\, dv_x = 0 \quad \text{[mean value]} \tag{2.17}$$

which vanishes because $P(v_x)$ is an even function on the integration interval and it is multiplied by an odd function, v_x.

The mean square of v_x (also called its variance) is given by

$$\boxed{\overline{(v_x - \overline{v}_x)^2} = \int_{-\infty}^{\infty} (v_x - \overline{v}_x)^2 P(v_x)\, dv_x} \quad \text{[variance]} \qquad (2.18)$$

and this can be shown to be

$$\overline{(v_x - \overline{v}_x)^2} = \sigma_{v_x}^2. \qquad (2.19)$$

Example 2.7 The escape velocity of a molecule is the least vertical velocity v_{esc} at which the molecule can escape the Earth's gravitational field. We can compute this velocity by finding the velocity a (collisionless) molecule might have upon falling from infinity to the Earth's surface. The procedure is to equate the kinetic energy of the particle $\frac{1}{2} m_0 v_{esc}^2$ to the potential energy at the Earth's surface GMm_0/R. After cancelling the m_0 on each side we find $v_{esc} = \sqrt{2gR}$ where $g = 9.8\,\mathrm{m\,s^{-2}}$ and $R \approx 6400\,\mathrm{km}$. The final answer is $v_{esc} = 11.2\,\mathrm{km\,s^{-1}}$, which is independent of mass (molecular species). □

There are many interesting pdf forms that arise in nature. These next two examples occur often. More cases can be found in elementary statistics books.

Example 2.8: uniform pdf

$$P(u) = \begin{cases} 1 & \text{if } 0 \le u \le 1 \\ 0 & \text{otherwise.} \end{cases}$$

After performing the integrals we find:

$$\mu_u = \frac{1}{2}, \quad \sigma_u^2 = \frac{1}{12}, \quad \sigma_u \approx 0.289. \qquad \square$$

Example 2.9: exponential distribution
This distribution is given by the formula

$$P(u) = \frac{1}{b} e^{-u/b} \qquad (2.20)$$

which has mean value b and variance b^2. □

We have already established the variance for an ideal gas from the relation (see (2.13))

$$\frac{1}{2} m_0 \overline{v_x^2} = \frac{1}{2} k_B T. \qquad (2.21)$$

Note that the factor of 3 seen before is not present because of the consideration of only the x component of velocity instead of all three components.

The three components of velocity are actually statistically independent of one another, and one of the rules of probability is that under these circumstances the *joint distribution* of the three variates is just the product of the individual densities:

$$\mathcal{P}(v_x, v_y, v_z) = \mathcal{P}(v_x)\mathcal{P}(v_y)\mathcal{P}(v_z) \tag{2.22}$$

$$= \left(\frac{m_0}{2\pi k_B T}\right)^{3/2} \exp\left(\frac{-m_0 v_x^2}{2k_B T} + \frac{-m_0 v_y^2}{2k_B T} + \frac{-m_0 v_z^2}{2k_B T}\right)$$

where $\exp(\cdot)$ stands for $e^{(\cdot)}$. And of course, the square of the velocity vector of a molecule is given by the sum of the squares of its components:

$$\boxed{v^2 = v_x^2 + v_y^2 + v_z^2} \quad \text{[speed from velocity]} \tag{2.23}$$

or written more compactly

$$\mathcal{P}(v_x, v_y, v_z)\, dv_x\, dv_y\, dv_z = \left(\frac{1}{2\pi\sigma^2}\right)^{3/2} \exp\left(\frac{-v^2}{2\sigma^2}\right) dv_x\, dv_y\, dv_z \tag{2.24}$$

where

$$\boxed{\sigma^2 = \frac{k_B T}{m_0}} \quad \text{[variance of velocity component]}. \tag{2.25}$$

The probability density function for molecular velocities (2.23) is called the *Maxwell–Boltzmann distribution* (see Table 2.4).

The distribution of velocities has no dependence on direction, only on speeds (i.e., it is *isotropic*). We can go to spherical coordinates in the velocity space and replace $dv_x\, dv_y\, dv_z$ by $v^2 \sin\theta\, d\theta\, d\phi\, dv$. Since there is no θ or ϕ dependence in the integrand we can integrate over them and the differential becomes $4\pi v^2\, dv$. The pdf (the remaining integrand) becomes a function of speed only:

$$\boxed{\tilde{\mathcal{P}}(v)\, dv = 4\pi v^2 \left(\frac{1}{2\pi\sigma^2}\right)^{3/2} \exp\left(\frac{-v^2}{2\sigma^2}\right) dv} \quad \text{[speed pdf]} \tag{2.26}$$

and the integrals now run from 0 to ∞. The last formula gives the probability of finding the speed of a molecule in the infinitesimal interval $(v, v + dv)$. A graph of this function is shown in Figure 2.6.

Table 2.4 *Comparison of characteristic velocity scales for a Maxwell–Boltzmann distribution; the values are for the "hypothetical" air molecule (M = 29)*

Quantity	Math form	Formula	Value at 300 K (m s^{-1})
rms velocity	$v_{rms} = (\overline{v^2})^{1/2}$	$\sqrt{\frac{3k_BT}{m_0}} = \sqrt{\frac{3R^*T}{M}} = \sqrt{3RT}$	508
mean speed	$\overline{v}$	$\sqrt{\frac{8k_BT}{\pi m_0}} = 0.921v_{rms}$	811
mode speed	v_m	$\sqrt{\frac{2k_BT}{m_0}} = 0.816v_{rms}$	415
speed of sound (air)	v_S	$\sqrt{\frac{7k_BT}{5m_0}} = 0.683v_{rms}$	331
standard deviation (air)	σ	$\sqrt{\frac{k_BT}{m_0}} = 0.577v_{rms}$	293

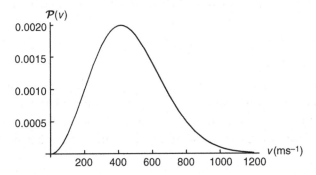

Figure 2.6 Graph of the distribution of molecular speeds $4\pi v^2 \mathcal{P}(v)$ for air molecules ($M = 29$) at a temperature of 300 K. The speed v is expressed in units of m s^{-1}. Recall that the escape velocity is 11.2 km s^{-1} independent of mass. The value of the probability density function at $v = 38\sigma$ is 10^{-313} s m^{-1}, which might help to explain why the Earth retains its atmosphere.

It is interesting to compare the escape velocity (11 200 m s^{-1}) with the distribution shown in Figure 2.6. The value of the distribution is some 10^{-313} s m^{-1}. The number of these molecules to escape even over the history of the planet (4.7×10^9 years) is exceedingly small. The median of the distribution moves to higher speeds if the temperature is raised or if the mass of the molecules is lowered. For example, Figure 2.7 shows the case of hydrogen atoms ($M = 1$) at 350 K, a value characteristic of altitudes ~ 120 km. The value of the density distribution at the escape velocity is 8×10^{-12} s m^{-1}, and after numerical integration of the

Gases

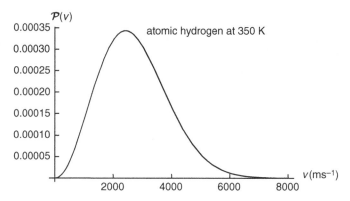

Figure 2.7 Graph of the distribution of molecular speeds $4\pi v^2 \mathcal{P}(v)$ for atomic hydrogen ($M = 1$) at a temperature of 350 K, a value for the upper atmosphere ($\sim$ 120 km). The speed v is expressed in units of m s^{-1}. Recall that the escape velocity is 11.2 km s^{-1} independent of mass. The value of the probability density function at $v = 11.2$ km s^{-1} is 8×10^{-12} s m^{-1}, and the area under the rest of the curve is 2×10^{-9}. This is a value large enough that if H reaches the upper atmosphere it will be depleted over planetary lifetimes. However, H is continually produced in the upper atmosphere by photodissociation (see Chapter 8) of water vapor.

density from the 11.2 km s^{-1} to infinity, we find that the probability of the speed being higher than the escape value is 2×10^{-9}. This is probably large enough for H to escape, but small enough that water molecules steadily being disassociated by hard (very short wave) solar radiation can maintain a presence at very high altitudes.

Example 2.10 By contrast the Moon has a smaller radius ($0.24\, r_E$ =1737 km) and mass (7.349×10^{22} kg $= 0.01229 M_E$) than Earth. This means that the acceleration of gravity is

$$g_{Moon} = GM_{Moon}/R^2_{Moon} = 1.63\,\text{m}\,\text{s}^{-2}$$

and

$$v_{esc} = \sqrt{2g_{Moon}R_{Moon}} = 2380\,\text{m}\,\text{s}^{-1}.$$

The maximum surface temperature on the Moon is about 400 K. Figure 2.8 shows the distribution of speeds for this case. The integral from the escape velocity to infinity is 2.73×10^{-7}, easily large enough for the Moon to lose its atmosphere over its lifetime. □

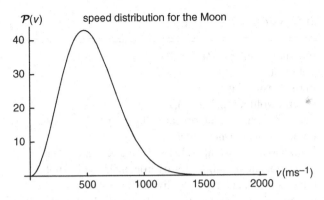

Figure 2.8 The distribution of molecular air molecules for the Moon at 400 K. The escape velocity is $2380\,\mathrm{m\,s^{-1}}$.

2.3 Flux of molecules striking a wall

There are many derivations of elementary processes in kinetic theory. We present one more here since the result comes up often. We want to know the number of molecules striking a wall (perpendicular to the x-axis) per unit time and area. This is simply the number density times the x component of velocity averaged over the velocity distribution. We proceed by finding $n_0\bar{v}_x$ using the Maxwell–Boltzmann distribution for the x component (the other factors for the y and z components integrate to unity). We consider only the positive component of v_x:

$$\text{flux/}(\perp \text{ area}) = n_0\overline{v_x} = n_0 \int_0^\infty A v_x \exp\left(-\frac{v_x^2}{2\sigma^2}\right) dv_x \qquad (2.27)$$

with $A = (m_0/(2\pi k_B T))^{1/2}$ and $\sigma^2 = k_B T/m_0$. The integral can be evaluated to give

$$\text{flux/}(\perp \text{ area}) = n_0\overline{v_x} = n_0 \left(\frac{k_B T}{2\pi m_0}\right)^{1/2}. \qquad (2.28)$$

And finally:

$$\boxed{\text{flux/}(\perp \text{ area}) = \frac{1}{4} n_0 \bar{v}} \quad \text{[flux of molecules hitting a wall]}. \qquad (2.29)$$

If we apply this to leaks through a small hole in the wall the process is called *effusion*. The formula holds when the hole is smaller than the mean free path of the molecules so that they flow through the hole without collisions; otherwise the gas acts like a fluid when passing through the opening and one must use fluid mechanics methods rather than kinetic theory.

Table 2.5 *Gas notation*

p	pressure ($N\,m^{-2} = Pa$; $100\,Pa = 1\,hPa = 1\,mb$)
V	volume (m^3)
ρ	mass density ($kg\,m^{-3}$)
α	specific volume ($m^3\,kg^{-1}$), $\alpha = \rho^{-1}$
m_0	mass of an individual molecule (kg); for H, $m_0 = 1.67 \times 10^{-27}\,kg$
n_0	number density (molecules m^{-3})
k_B	Boltzmann's constant: $1.381 \times 10^{-23}\,J\,K^{-1}\,molecule^{-1}$
M_G	gram molecular weight; for hydrogen, $M_H = 1\,g\,mol^{-1}$
$\tilde{M}_G$	the gram molecular weight divided by 1000
M_d	dry air effective molecular weight, $M_d = 28.97\,g\,mol^{-1}$
M_E	effective gram molecular weight of a mixture of gases
$\mathcal{M}_i$	bulk mass of constituent i (kg)
N_A	Avogadro's number: 6.022×10^{23} molecules mol^{-1}
ν	number of moles of a gas
R^*	universal gas constant: $8.3143\,J\,K^{-1}\,mol^{-1}$
R_d	gas constant for dry air: $287\,J\,K^{-1}\,kg^{-1}$
R, R_G	gas constant for a particular gas, $G\,(J\,K^{-1}\,kg^{-1})$

2.4 Moles, etc.

The *molecular weight*, M, of a pure gas is the sum of the atomic weights of the atoms making up the molecules. The molecular weight has dimensions grams per mole, denoted $g\,mol^{-1}$ (see Table 2.5). In keeping with SI units one might choose $kg\,kmol^{-1}$, which gives the same numerical value. For example, the molecular weight of isotopically pure (no deuterium (^2H) or tritium (^3H) atoms in the gas) H_2 is 2 and that of CO_2 is $12 + 16 + 16 = 44$. The chemical properties of the element are determined by the number of protons in the nucleus, which is designated the *atomic number*. The *atomic weight* is determined by the sum of the number of protons and the number of neutrons. An element can have different *isotopes*, i.e., the number of neutrons might vary slightly from atom to atom. But the most abundant isotope found in nature is usually dominant, with only a small percentage of the other isotopes present. If we take a random sample from nature this leads to a weighted average of the atomic weight, and this is the value used in most computations. For our purposes, we can simply use the numbers given in Table 2.6 which take into account the distributions of naturally occurring isotopes. Strictly speaking the standard is set by the most abundant isotope of carbon which is defined to have a molecular weight of exactly $12.000\,kg\,kmol^{-1}$.

The number of molecules in a gram mole is called *Avogadro's number*

$$\boxed{N_A = 6.022 \times 10^{23} \text{ molecules } mol^{-1}} \quad \text{[Avogadro's number].} \qquad (2.30)$$

Table 2.6 *Selected atomic and molecular weights in grams per mole (equivalent numerically to kg kmol^{-1})*

Ar	39.948	N	14.0067
N_2	28.0134	S	32.066
O_2	31.9988	H_2O	18.01534
CH_4	16.04303	dry air	28.97
C	12.01115	H	1.00794
H_2	2.01594	He	4.0026
O	15.9994		

The number of moles of a substance is denoted by ν. It is the number of molecules N under consideration divided by N_A:

$$\boxed{\nu = \frac{N}{N_A}} \quad \text{[number of moles].} \tag{2.31}$$

In terms of number density ($n_0 = N/V$), the number of moles is given as $\nu = n_0 V/N_A$, where V is the volume occupied by the gas in m^3. The mass (in grams!) of an individual molecule is related to these quantities by $m_0 = M_G/N_A$, where M_G is known as the (gram-)molecular weight of the gas G. (Avogadro's number of pure hydrogen atoms has a mass of 1g.) The gram-molecular weight is the one used in most existing tables; hence, we use it here. But in keeping with our SI units, we need to express m_0 in kilograms in most formulas. Thus the applicable expression is

$$\boxed{m_0 = \frac{M_G}{N_A}} \quad \text{[m_0 in grams].} \tag{2.32}$$

However,

$$m_0 = \frac{1}{N_A}\tilde{M}_G, \quad \text{with } \tilde{M}_G = \frac{M_G}{1000} \quad \text{[m_0 in kilograms].} \tag{2.33}$$

The Ideal Gas Law can then be written:

$$p = \rho R_G T \tag{2.34}$$

$$= n_0 m_0 R_G T \ (\rho \text{ in kg m}^{-3}, p \text{ in N m}^{-2}, m_0 \text{ in kg})$$

since $\rho = n_0 m_0$. Now $n_0 = \nu N_A/V$, the total number of molecules per unit volume. Also note that m_0 (in kilograms) is $\tilde{M}_G/N_A$. Then

$$p = \frac{\nu N_A}{V}\frac{1}{N_A}\tilde{M}_G R_G T \tag{2.35}$$

$$p = \frac{\nu}{V}R^* T \tag{2.36}$$

where

$$\boxed{R^* = \tilde{M}_G\,R_G}\quad [\text{relating } R^* \text{ and } R_G]\tag{2.37}$$

is called the *universal gas constant* ($R^* = 8.3145\,\mathrm{J\,K^{-1}\,mol^{-1}}$).
 Note the relationship:

$$\boxed{R^* = k_B N_A}\quad [\text{relating } R^* \text{ and } k_B].\tag{2.38}$$

In other words, the Boltzmann constant is the gas constant per molecule while R^* is the gas constant for a mole of molecules. The gas constant for a specific gas is related to the universal gas constant by

$$R_G = \frac{R^*}{\tilde{M}_G}\quad [\text{in J K}^{-1}\,\text{kg}^{-1}]\tag{2.39}$$

$$= \frac{k_B N_A}{\tilde{M}_G}\tag{2.40}$$

$$= \frac{k_B}{m_0}\quad [m_0 \text{ in kg}].\tag{2.41}$$

Again the use of $\tilde{M}_G$ instead of M_G above simply converts to SI units; this is because of the conventional definition of the mole in terms of grams instead of kilograms.

Standard conditions (denoted STP) for a gas are $0\,^\circ\mathrm{C}$ and 1 atm of pressure ($= 1013\,\mathrm{hPa}$). It is useful to know that 1 mol of an ideal gas occupies 22.4 liters ($1\,\text{liter} = 1000\,\mathrm{cm}^3 = 10^{-3}\,\mathrm{m}^3$) or $0.0224\,\mathrm{m}^{-3}$ at STP.

Example 2.11 Calculate the gas constant for dry air.
Answer: Use (2.39) with $M_d = 28.97$. $R_d = 287\,\mathrm{J\,kg^{-1}\,K^{-1}}$.   □

Example 2.12 Calculate the gas constant for water vapor.
Answer: Same as above, $M_w = 18.015$, $R_w = 461.5\,\mathrm{J\,kg^{-1}\,K^{-1}}$.   □

Example 2.13 Calculate the densities of pure dry air and pure water vapor at STP.
Answer: Use the gas law, $\rho = p/R_G T$:

$$\rho_d(\text{STP}) = 1.013 \times 10^5\,\mathrm{Pa}/(287\,\mathrm{J\,kg^{-1}\,K^{-1}})(273.2\,\mathrm{K}) = 1.292\,\mathrm{kg\,m^{-3}}\tag{2.42}$$

$$\rho_w(\text{STP}) = 1.013 \times 10^5\,\mathrm{Pa}/(461.5\,\mathrm{J\,kg^{-1}\,K^{-1}})(273.2\,\mathrm{K}) = 0.803\,\mathrm{kg\,m^{-3}}.$$
$$\tag{2.43}$$

Water vapor at the same temperature and pressure is less dense than dry air.   □

Example 2.14 A vessel contains $1.2\,\mathrm{kg}$ of dry air at STP. How many moles of O_2, N_2, and Ar are there?

Answer: $1200\,\text{g} = v_\text{d} M_\text{d}$. Then $v_\text{d} = 1200/28.97 = 41.4$ mol of dry air. We know that $v_{O_2} = 0.21 v_\text{d}$, $v_{N_2} = 0.78 v_\text{d}$, $v_\text{Ar} = 0.0093 v_\text{d}$. Inserting, we find: $v_{O_2} = 8.70$ mol, $v_{N_2} = 32.3$ mol, and $v_\text{Ar} = 0.41$ mol. □

Example 2.15 How high would a molecule with upward directed speed $485\,\text{m s}^{-1}$ go before turning back in the Earth's gravity field?
Answer: From elementary mechanics the height of such a flight is given by converting all the initial kinetic energy into potential energy:

$$\frac{1}{2}mv_0^2 = mgh \qquad (2.44)$$

or $h = v_0^2/(2g) \approx 12000\,\text{m} = 12\,\text{km}$. This is just larger than the scale height of the atmosphere. □

Example 2.16 What is the average force exerted by a molecule of mass m_0 making elastic reflections on the floor under the influence of gravity?
Answer: Let the molecule fall from a height h. The time for its fall is $t = \sqrt{2h/g}$. Its speed upon impact is $v_0 = \sqrt{2gh}$. The momentum change on a reflection is $2m_0 v_0 = 2m_0\sqrt{2gh}$. The force exerted on the floor is then $2m_0 v_0 / \Delta t$; Δt is the time for a round trip up and back down by the molecule. We finally obtain,

$$F = \frac{2m_0\sqrt{2gh}}{2\sqrt{2h/g}} = m_0 g. \qquad (2.45)$$

In other words, the average force on the floor is just the weight of the bouncing molecule. Note that the result is independent of the initial dropping height h. Is it any wonder that atmospheric pressure measures the weight of air in the column above a square meter? □

2.5 Dalton's Law

Dalton's Law deals with a mixture of ideal gases. It states that the *partial pressures* of the individual constituent gaseous components are additive. Based upon the kinetic theory derivation above it is not surprising that the pressures would be additive for mixtures of ideal gases with different molecular masses $m_1, m_2, \ldots$.
Writing the expression for the sum of partial pressures,

$$p = p_1 + p_2 + \cdots. \qquad (2.46)$$

Once we accept this rule we can compute the effective gas constant for a mixture of gases. To do so we substitute for the partial pressure of the different constituents:

$$p = (\rho_1 R_1 + \rho_2 R_2 + \cdots)T. \qquad (2.47)$$

After multiplying and dividing through by ρ, the mass density of the mixture,

$$p = \rho \left(\frac{\rho_1}{\rho} R_1 + \frac{\rho_2}{\rho} R_2 + \cdots \right) T. \tag{2.48}$$

After noting that the volumes are the same for ρ_i and ρ, we can write

$$\boxed{R_{\text{eff}} = \frac{\mathcal{M}_1}{\mathcal{M}} R_1 + \frac{\mathcal{M}_2}{\mathcal{M}} R_2 + \cdots} \quad \text{[effective gas constant for a mixture].} \tag{2.49}$$

Note that the coefficients in the last equation are mass fractions, not mole fractions.

It might be necessary to calculate the *effective molecular weight* of a mixture (as we have used $M_d = 29 \, \text{g mol}^{-1}$ for dry air). We can start with

$$R^* = k_B N_A, \quad R = \frac{k_B}{m_0} = \frac{k_B N_A}{m_0 N_A} = \frac{R^*}{M}, \tag{2.50}$$

where M is the molecular weight (kg mol^{-1}). Now return to (2.49) and set

$$R_{\text{eff}} = \frac{R^*}{M_{\text{eff}}}. \tag{2.51}$$

This leads to

$$\boxed{M_{\text{eff}} = \frac{R^*}{R_{\text{eff}}}} \quad \text{[effective molecular weight for a mixture].} \tag{2.52}$$

As a check let $R_{\text{eff}} = 287 \, \text{J kg}^{-1} \, \text{K}^{-1}$, and we find $M_{\text{eff}} = 0.029 \, \text{kg mol}^{-1} = 29 \, \text{g mol}^{-1}$.

Example 2.17 An argon atmosphere. What is R_{Ar}?
Answer: $R_{\text{Ar}} = R^*/\tilde{M}_{\text{Ar}} = 8.31 \times 10^3 \, \text{J K}^{-1} \, \text{kmol}^{-1} \, / \, 39.94 \, \text{kg kmol}^{-1} = 208.2 \, \text{J K}^{-1} \, \text{kg}^{-1}$.
□

Example 2.18 What is the density of argon gas at STP?
Answer: $\rho = p/(R_{\text{Ar}} T) = (101325 \, \text{Pa})/(208.2 \, \text{J K}^{-1} \text{kg}^{-1} \times 273 \, \text{K}) = 1.78 \, \text{kg m}^{-3}$.
□

Example 2.19 Suppose the atmospheric density is given by $\rho(z) = \rho_0 e^{-z/H}$, where z is altitude above sea level, $\rho_0 = 1.2 \, \text{kg m}^{-3}$ and H is the scale height, nominally 10 km. What is the mass of air above $1 \, \text{m}^2$ at sea level?
Answer: dm for a slab of thickness dz is $\rho(z)A \, dz$ where A is the horizontal cross-sectional area of the slab. Adding up all the infinitesimal slabs in the column amounts to

$$\mathcal{M} = \int_0^\infty A\rho_0 \, e^{-z/H} \, dz = A\rho_0 H = 1 \, \text{m}^2 \times 1.2 \, \text{kg m}^{-3} \times 10^4 \, \text{m} = 1.2 \times 10^4 \, \text{kg}.$$

□

Notes

There are many good classical thermodynamics books. One which is very readable and includes a mix of elementary kinetic theory as well is Sears (1953). Another more recent but very readable account is by Houston (2001). A purely thermodynamic treatment is given in the physics text by Zemanski (1968). Modern physical chemistry books are perhaps best for discussions of gas thermodynamics, for example Atkins (1994). For a readable but rigorous discussion of constraints, etc., see Reiss (1965).

Notation and abbreviations for Chapter 2

a	Bohr radius
F	force (N)
h	Planck's constant (J s)
h_0	initial height
k_B	Boltzmann's constant (Table 2.5)
L	length of box edge (m)
λ	mean free path (m)
m_e	electron mass (kg)
m_0	mass of a single molecule (kg)
M_{eff}	effective molecular weight ($g\,mol^{-1}$)
M_G	gram molecular weight of a gas ($g\,mol^{-1}$)
$\mathcal{M}$	bulk mass of an object (kg)
$n_0(z)$	molecular density as a function of height (molecules m^{-3})
N	total number of molecules
N	newtons
N_A	Avogadro's number
p, p_G	pressure, partial pressure for gas G (Pa)
$\mathcal{P}(v_x, v_y, v_z)$	joint probability density function for velocity components
$\mathcal{P}(z)$	probability density function
r_0	effective molecular radius (m, nm)
R^*	universal gas constant ($J\,mol^{-1}\,K^{-1}$)
R_d	gas constant for dry air ($J\,kg^{-1}\,K^{-1}$)
R_{eff}	effective gas constant for a mixture of gases ($J\,kg^{-1}\,K^{-1}$)
R_w	gas constant for water vapor ($J\,kg^{-1}\,K^{-1}$)
ρ	density ($kg\,m^{-3}$)
Δt	time interval (s)
T	temperature (K)
σ	standard deviation

σ^2 variance
σ_c collision cross-section (m^2)
(u, v, w) velocity components (v_x, v_y, v_z)
v speed ($m\,s^{-1}$)
$\overline{v}$ mean speed ($m\,s^{-1}$)
v_{esc} escape velocity ($m\,s^{-1}$)
$\overline{v^2}$ mean square velocity ($m^2\,s^{-2}$)
$\overline{v_x^2}$ mean square of x component of velocity ($m^2\,s^{-2}$)
V volume (m^3)
z elevation (m)

Problems

2.1 Calculate the mass of a 1 m^3 parcel of dry air at STP. Calculate its mass at the same pressure but at 10 °C and 20 °C.

2.2 Calculate the mass of a 1 m^3 parcel of water vapor at STP.

2.3 What is the partial pressure of oxygen in a dry 1 m cube of air at STP?

2.4 What is the weight of the 1 m cube of dry air at STP? In newtons, in pounds? (Note: 1 kg weighs 2.2 lb at sea level.)

2.5 What is the number density of a volume of pure oxygen (O_2) at STP?

2.6 Express R_d in terms of hPa instead of Pa.

2.7 Use $d\mathcal{P}(v)/dv = 0$ to find a formula for the most probable speed of a molecule at STP.

2.8 Compute the v_{rms} for O_2, N_2, Ar, and H_2 at STP. Compare to the escape velocity.

2.9 The mass of a certain air parcel is 1 kg, its temperature is 0 °C and it occupies a volume of 1 m^3. It is known to have 5 g of water vapor and the rest is dry air. What is the partial pressure of water vapor? What is the density of this moist air? Compare to the density of dry air at the same overall pressure.

2.10 A cylindrical column of air has radius 1 km. The surface pressure is 1000 hPa. The entire column is rising at a speed of 10 cm s^{-1}. What is the kinetic energy of the column?

2.11 The cylinder of the previous problem is rotating about its axis of symmetry at a rate of 2π radians per day (1 day = 8.64×10^4 s). What is its rotational kinetic energy? (*Hint:* The moment of inertia of a cylindrical slab is $I = \frac{1}{2}mR^2$; kinetic energy = $\frac{1}{2}I\omega^2$ where ω is angular velocity in rad s^{-1}.)

2.12 Suppose the number density of molecules in a column of air is given by $n_0(z) = n_0(0)e^{-z/H}$. What is the total number of molecules in a column with unit cross-sectional area? What are reasonable values for $n_0(0)$ and H? Use STP at $z = 0$.

2.13 Given the conditions of the last problem and a reasonable value for σ_c what is the approximate altitude z_H for which the mean free path, λ, is equal to H?

2.14 Given that the molecules in a column of air are distributed vertically as in the last two problems and that the temperature is constant in the column at T_0, what is the total gravitational potential energy in the column in $J\,m^{-2}$?

2.15 Suppose the column of air in the last problem is isothermal at a temperature of 300 K. What is the total kinetic energy of the molecules in the column?

2.16 Compare the collision frequency of "air" molecules at STP to the frequency of yellow light. Note also for comparison that the lifetime of an excited state of an atom is about 10^{-8} s.

3

The First Law of Thermodynamics

Often in meteorology we deal with a fixed mass of a gas whose volume and other characteristics may change as the air mass moves about. The particular mass of gas may be thought of as a small parcel of matter that is transported through the environment by natural forces acting upon it. We could also imagine moving it virtually via an abstract thought experiment, for example to determine its stability under small perturbations. As an air parcel rises for whatever reason in the real atmosphere, it will almost instantaneously adapt its internal pressure to the external pressure exerted by the local surroundings, but the temperature and composition adjust more slowly. In convection, entrainment of neighboring air also speeds up the process of equalizing the temperature between inside and outside air. Still there is a huge separation of equalization times between pressure and temperature and/or trace gas concentrations. This time scale separation has made the parcel concept a useful and even powerful tool in the atmospheric sciences. We will return to it often.

Thermodynamics is concerned with the *state of a system* (an example of which is the parcel alluded to above now treated as an approximate thermodynamic system) and the changes that occur in its state when certain processes occur (such as its being lifted). In the case of a parcel composed of an ideal gas, the state is completely described by the state variables p, V, $\mathcal{M}$ and T (actually in equilibrium only three variables need to be specified, since the equation of state in the form of the Ideal Gas Law can be used to calculate the fourth from knowledge of the other three). In practice in our applications using parcels, the total mass $\mathcal{M}(\mathcal{M} = \sum_i \mathcal{M}_i$ where i indicates different species such as O_2, N_2, etc.) is usually also held fixed (unless otherwise specified), making only three state variables. There are some other state variables more directly related to energetics that are convenient for certain purposes and much of this chapter will be concerned with the first two of them, *internal energy* and *enthalpy*.

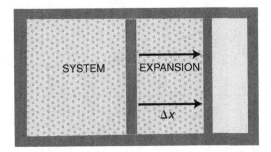

Figure 3.1 Illustration of an expansion of the system in the x direction by a distance Δx. The force exerted by the system on the movable wall is pA, where A is the area of the wall. The work done can then be expressed as $pA\,\Delta x = p\,\Delta V$ where ΔV is the change in volume of the system.

The most basic energetic quantity to consider is the *work* performed by the gas on its surroundings during an expansion of the system's (e.g., the parcel's) volume (see Figure 3.1). If the system (whose shape we take to be the volume defined by the cylinder of Figure 3.1) expands in the x direction by a distance Δx, then the force exerted *on* the environment is $F = pA$, where A is the area of the movable wall in the cylinder and p is the pressure the system exerts on the wall. The amount of work done by the parcel *on* the environment in this infinitesimal process is [1]

$$\text{work done by the system} = F\,\Delta x = pA\,\Delta x = p\,\Delta V$$

where ΔV is the infinitesimal change in volume associated with the expansion in the x direction. The expansion need not be solely along the x-axis as shown in the figure, but can be a distortion in any or all directions. The formula $W = p\Delta V$ still holds for infinitesimal ΔV.

If the pressure is not constant during the expansion, we must sum the contributions from a sequence of infinitesimal expansions

$$W_{V_A \to V_B} = \int_{V_A}^{V_B} p\,dV \quad \text{[expansion work done by the system].} \quad (3.1)$$

On a coordinate plane with pressure and volume as ordinate and abscissa, the work performed is the area under the curve of p versus V (Figure 3.2). This makes it clear that the amount of work done by the system on its environment depends on the path taken $p = p(V)$, which defines a curve in the plane. One might imagine a cyclic process in which the system expands from V_A to V_B along one path and returns along another path. The area between the two curves represents the net work

---

[1] In this text we use the sign convention that positive W means work done *by* the system on its environment. Some textbooks use the opposite sign for work done by the system.

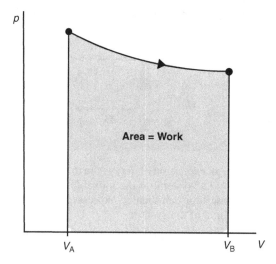

Figure 3.2 Graph of pressure p versus volume V for an expansion of a gaseous system from V_A to V_B along a specified curve $p(V)$. The area under the curve is the work performed by the system on the environment.

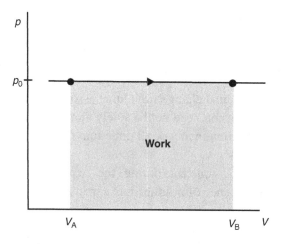

Figure 3.3 Graph of pressure p versus volume V for an expansion of a gaseous system from V_A to V_B along a constant pressure path (isobaric path). The area under the curve is the work performed by the system on the environment.

done by the parcel on its surroundings during the cyclic process. When the parcel's volume decreases, it does negative work.

Example 3.1 Consider a system composed of 2 kg of dry air at 0 °C. Let the system expand isobarically from its initial volume of 1 m³ to 3 m³. How much work is done by the system? (See Figure 3.3.)

Answer: The amount of work done by the gaseous system on its environment is given by the formula

$$W = \int_{V_A}^{V_B} p_A \, dV = p_A \int_{V_A}^{V_B} dV = p_A(V_B - V_A) \tag{3.2}$$

where $V_A = 1 \, m^3$ and $V_B = 3 \, m^3$, $p_A = \rho R_d T_A$. The density is given by $2 \, kg/1 \, m^3$, $T_A = 273 \, K$, $R_d = 287 \, J \, kg^{-1} \, K^{-1}$; hence $p_A = 1.57 \times 10^5$ Pa. Finally, $W = 3.13 \times 10^5 \, J = 313 \, kJ$ (kilojoules). □

Example 3.2 How much does the temperature change during the above process? *Answer*: $T_B = p_A/(\rho_B R_d)$. We can compute $\rho_B = mass/V_B = 2 \, kg/3 \, m^3 = 0.667$ $kg \, m^{-3}$. Then $T_B = 820 \, K$. Obviously an isobaric process leading to a tripling of volume is very unlikely in the atmosphere. □

3.1 Reversible and irreversible work

In the preceding we assumed that the work done by the system was along a well-defined path $p(V)$. Actually this is a rather strong assumption – that at each infinitesimal adjustment the curve $p(V)$ exists. We are implicitly assuming that we are in a state of thermodynamic equilibrium at each step – in other words the system has time to come to equilibrium (i.e., uniform temperature throughout, etc.) before the next infinitesimal step. In real processes such as the compression of a piston in an internal combustion engine, the gas in the chamber might be highly nonuniform and locally disturbed by such things as shock waves during the next change in volume (perhaps the equation of state does not even hold during this interval of time). For an *irreversible change* such as in the internal combustion engine, an amount of work will be done, but it may not be calculable using $\int p \, dV$. In more advanced books on thermodynamics it is shown that when the system does work (for example by expansion) $\int p \, dV$ is the *maximum work* that can be done. But when the system is compressed, the reversible (calculable) work ($\int p \, dV$) is the minimal work done by the system during the compression. In the high compression engine the amount of work done is seldom more than 75% of the estimate based on the reversible assumption. The unfortunate mechanical engineer simply cannot win in the face of irreversible processes. Luckily, most natural processes of interest to the atmospheric scientist are better behaved.

The idealization of *reversible work* allows us to do calculations using $\int p \, dV$ even though in reality it never quite works that way. In most applications that follow in this book the assumption of reversible work is adequate.

3.2 Heating a system

In the expansion process undergone by a parcel (fixed mass) in moving from (V_A, p_A) to (V_B, p_B), the parcel will do some work on the environment, $\int_{V_A}^{V_B} p(V) \, dV$. During this process some energy might be transported into the system because of a temperature difference between the interior of the system and the surrounding environment. This energy transported into the system is *thermal energy* [2] as described in Chapter 2. Thermal energy consists of all the modes of energy associated with individual molecules: translational kinetic energy, rotational energy for polyatomic molecules and vibrational energy (potential and kinetic) for polyatomic molecules that experience internal stretching oscillations. These individual energy terms each contribute to the thermal energy of a molecule (but only the translational kinetic energy contributes to pressure). In fact, the energy on the average is shared equally between the different modes (translational kinetic, etc.), although at atmospheric temperatures the vibrational modes are not excited because of quantum threshold effects. [3]

This transport of heat is effected at the molecular level by the collisions of individual molecules. If there is a gradient of temperature, molecules from the warmer region will penetrate a distance of the order of a mean free path before suffering a collision into the cooler region (and vice versa), causing the cooler region to warm; molecules moving the other way cause the warmer region to cool through individual collisions. The random motions of molecules crossing the boundary bring the news of their different "temperature" via a *random walk* process (each step forward or backward determined by the proverbial flip of a coin). The news and conversion are brought about slowly but surely. The distance advanced by the spreading edge of a "warm front" at the molecular level is proportional to the square root of the time elapsed. This is in stark contrast to the propagation of pressure differences which move via a sound-like wave (distance of advance of the pressure front being proportional to time). To obtain an idea of this contrast consider a one-dimensional gas (x direction only) and let an instantaneous hot spot develop at $x = 0$ (perhaps a fire cracker explodes). It is possible to solve this problem analytically, but the details need not be given here. The basic idea is that heat flux

---

[2] In most texts this thermal motion is referred to as *heat* as though it were a material substance moving around in space, but some authors (e.g., Bohren and Albrecht, 1998) shun the use of the noun *heat* in favor of the verbs *heating* or *cooling* as a transport process involving the energizing of neighboring molecules by their aggregate being in contact with an aggregate of molecules of a different temperature. We will use the term *heat* to mean the integral over the heating rate with respect to time. Just keep in mind that heat is not a fluid flowing about in the medium.

[3] The energy levels in quantum mechanics are discrete and the disturbing collisions need to have a sufficient energy transfer to effect a transition to the next higher energy level. Typically, rotational levels are closely enough spaced for them to be excited, but vibrational thresholds are much higher, requiring very high temperatures for excitation.

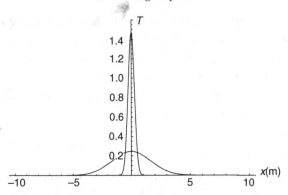

Figure 3.4 Spread in meters of a localized pulse of thermal energy due (only) to molecular thermal conductivity in air at STP after 2 h and after 12 h. The functional form is the normal distribution with standard deviation $\sigma = \sqrt{2Dt}$, where $D = \kappa_H/\rho c_p \approx 2 \times 10^{-5}$ m^2 s^{-1} and t is time in seconds.

is proportional to $-\kappa_H \mathrm{d}T/\mathrm{d}x$, where κ_H is the *thermal conductivity* of air (≈ 0.024 J m^{-1} s^{-1} K^{-1}). The pulse spreads out in the shape of a normal distribution as shown in Figure 3.4. The standard deviation of the spread of the elevated temperatures is only about 3 m after 12 hours. This means that the concept of parcel integrity for objects of the order of several hundred meters is safe for days if the only stirring mechanism is molecular diffusion. There are other mechanisms that can shorten the time of mixing depending on the conditions, but these are still usually slow compared to the adjustment of the interior to the exterior pressure. Note that sound waves travel at several hundred meters per second. The adjustment of pressures should be accomplished in several hundred passes of sound waves back and forth across the parcel – still very fast compared to molecular and even eddy (turbulent) transport processes. The sound waves are eventually dissipated into thermal energy.

Thermal energy or heat as we have been discussing it can now be contrasted with the *work* being done by a system during a process. The thermal energy is at the molecular level and it migrates from place to place via gradients in the temperature (say from the system to a reservoir), while work is at the truly macroscopic level. Work is performed when one of the macroscopic dimensions (sometimes called *configuration coordinates*), say the position of a piston, is altered a finite (macroscopic) amount.

Returning to our system, heat can be transported into it because of small differences in temperature between the system and its environment. The amount of heat taken into the system during a finite process is traditionally given the symbol Q. We say [4] $Q_{A\rightarrow B}$ as the system moves from the state denoted A to the state denoted B.

---

[4] Note that in our notation if a positive amount of heat is *absorbed* by the system then $Q_{A\rightarrow B}$ is assigned a positive value. This is the sign convention followed by virtually all textbooks.

Consider first the simple heating of a parcel where the volume is held fixed (an *isochoric* process). The parcel is heated by an amount $Q_{A\rightarrow B}$ and its temperature undergoes a change $\Delta T = T_B - T_A$. The heat energy absorbed and the temperature change are related with the coefficient of proportionality being the mass times the specific heat capacity c_v (units $J\,kg^{-1}\,K^{-1}$); $C_v = \mathcal{M}c_v$, where C_v (units $J\,K^{-1}$) is called the total heat capacity:

$$\boxed{Q_{A\rightarrow B} = \mathcal{M}c_v\Delta T} \qquad \text{[heating at constant volume].} \qquad (3.3)$$

In general, c_v might be a function of temperature, but for an ideal gas it is not. For dry air the value of c_v is $717\,J\,kg^{-1}\,K^{-1}$. In this process no work is done by the parcel on its environment, since the volume of the parcel does not change. All the heat given to the system goes into its *internal energy*. From the molecular relation $\frac{1}{2}mv^2 = \frac{3}{2}k_B T$ we recall that the average kinetic energy of the molecules is proportional to the Kelvin temperature. Hence, a change in internal energy is equivalent to a change in the kinetic energy (for an ideal monatomic gas). As we will see later the kinetic energy of translation still has the same relation to temperature for multi-atomic gases, but the internal energy in the multi-atomic case is modified (next section).

In chemical applications (also chemical texts and handbooks) it is common to use the molar specific heat for a substance, $\bar{c}_v$. In this case the total heat capacity is $C_v = \nu\bar{c}_v$ where ν is the number of moles in the system. In this formulation:

$$Q = \nu\bar{c}_v\Delta T. \qquad (3.4)$$

Example 3.3 A sealed room with walls made of perfectly insulating material has dimensions $4\,m\times4\,m\times3\,m$. The conditions are $p = 1000\,hPa$, $T = 300\,K$. What is the mass of air in the room? How many joules are required to raise the temperature by $1\,K$?
Answer: $\mathcal{M} = pV/R_d T = (10^5\,Pa \times 48\,m^3)/(287\,J\,kg^{-1}\,K^{-1} \times 300\,K) = 55.7\,kg$.
$Q = c_v\mathcal{M}\Delta T = 717\,J\,kg^{-1}\,K^{-1} \times 55.7\,kg \times 1\,K = 4.0\times10^5\,J = 400\,kJ$. $\square$

Example 3.4 How many kilowatt hours of energy are expended in the last example?
Answer: $1\,kWh = 3600\,kJ$. So the result is $0.11\,kWh$. Note that the cost of $1\,kWh$ is a few cents (US). $\square$

Example 3.5 In the last example, consider the effect of thin walls. Let us take the walls to be wood and $1\,cm$ thick. What is the amount of heat necessary to bring these walls (and floor) up $1°C$?
Answer: The specific heat of wood is $1760\,J\,kg^{-1}\,K^{-1}$, and the density of wood is $600\,kg\,m^{-3}$. The walls, ceiling and floor have an area of $72\,m^2$ making a volume of $0.72\,m^3$. The mass of this material is $600\,kg\,m^{-3} \times 0.72\,m^3 = 432\,kg$. The total heat capacity of the solid matter is $760\,kJ$, nearly twice that of the air contained. $\square$

3.2.1 Internal energy and the First Law

When a system undergoes a transition from one state to another, energy passes from the system to its environment or vice versa, since energy in all its forms is conserved (even if the process is irreversible). The change in internal energy, ΔU, for a system is defined to be

$$\boxed{\Delta U = Q_{A \to B} - W_{A \to B}} \quad \text{[internal energy]}. \tag{3.5}$$

The differential form is

$$\boxed{dU = đQ - đW} \quad \text{[First Law]} \tag{3.6}$$

where the bar crossing the d of $đQ$ and $đW$ is to remind us that both of these differentials depend on the path (it is not a perfect differential as in multivariable calculus, whereas dU is). In the last expression dU is the infinitesimal change in *internal energy*. The last equation is a statement of the conservation of energy. The First Law of Thermodynamics actually goes much further and states that the internal energy U is a function only of the state of the system. For the ideal monatomic gas this is obvious from our simple kinetic theory model since the internal energy is the total kinetic energy summed over all the molecules in the system and this is proportional to the Kelvin temperature. For a given mass, being a function of the state means its value is uniquely determined at each point in a $V - p$ diagram. Note that neither Q nor W are functions of state.

3.3 Ideal gas results

3.3.1 Internal energy of an ideal gas

In addition to the ideal gas equation of state another property is necessary to define an ideal gas. We must specify its internal energy as a function of the thermodynamic coordinates. This can be accomplished by laboratory experiments to yield:

$$\boxed{U = \frac{f}{2} N k_B T} \quad \text{[internal energy of ideal gas]} \tag{3.7}$$

where N is the number of molecules in the system. There are many alternative forms of this relation because of the different ways we can describe an ideal gas:

$$U = \frac{f}{2} n_0 k_B T V = \frac{f}{2} \rho R T V = \frac{f}{2} p V \tag{3.8}$$

$$U = \frac{f}{2} M R T = \frac{f}{2} \nu R^* T \tag{3.9}$$

where k_B is Boltzmann's constant (Table 1.1), $\mathcal{M}$ is the mass of gas in the system, and f is a constant equal to 3 for an ideal monatomic gas (e.g., Ar), and $f = 5$ for most diatomic gases at room temperature (e.g., air). The internal energy can be determined by a series of experiments involving adiabatic processes in which $\Delta U = -\mathcal{W}_{A \to B}$ is easily measured. The constant f depends upon the internal structure of the molecules; it is known as the *number of degrees of freedom* in the molecule. For a diatomic molecule that is very stiff (does not stretch and contract under the temperature conditions being considered) such as O_2 and N_2 near room temperature, the value of f is 5 because there are two more degrees of freedom due to the ability of the molecule to rotate in a two-dimensional plane, but not about the axis joining the two atomic constituents (its moment of inertia is too small about the axis joining the two atoms). At high temperatures (not naturally occurring in the lower atmosphere) the number of degrees of freedom goes up by two because of molecular vibration due to the spring-like binding. Hence, for normal atmospheric conditions such as encountered in the troposphere and stratosphere:

$$f = 3 \quad \text{ideal monatomic gas (e.g., argon)} \tag{3.10}$$

$$f = 5 \quad N_2, O_2 \text{ near STP} \tag{3.11}$$

$$f = 6 \quad CO_2, H_2O \text{ and other nonlinear molecules.} \tag{3.12}$$

It is not difficult to see why the specific heat should be larger for molecules with larger f. Consider the constant volume heating case. If heating occurs in a box of monotonic gas, all the energy must go into increasing the linear (translational) kinetic energy of the molecules. If the molecule is spatially extended such as a diatomic molecule, it can rotate as well as translate. The added energy can go into rotational energy as well as translational energy. Hence, the heat capacity (amount of heat necessary to raise the system's temperature by 1 K) will be larger. Basically the heat energy (that at the molecular level) must be shared among all the degrees of freedom, but only the linear kinetic energy goes into causing pressure since it carries momentum to the walls (or across boundaries).

A remarkable theorem proved in the classical study of statistical mechanics shows that in equilibrium the energy will be shared equally between each of the rotational and translational modes (and vibrational modes when applicable): *the principle of equipartition of energy*. In the case of a diatomic molecule only two rotational modes are available, the rotation about the axis joining the two atoms does not count. In the case of a triatomic molecule in which the atoms are not in a straight line (e.g., CO_2), all three rotational modes are involved: $f = 6$. Molecules actually can vibrate (stretching and contracting like masses joined by a spring) as well, and at sufficiently high temperatures these modes can enter and raise f even more, but as remarked above, this vibrational degree of freedom is not important in most applications of thermodynamics to the atmosphere. On the other hand, such modes of vibration and rotation play an important role in the absorption and emission of infrared radiation as it passes through air. In summary, each molecule on the average possesses $\frac{1}{2}k_BT$ for each of its mechanical degrees of freedom. For 1 mol of such molecules at 300 K this is $\frac{1}{2}R^*T = 2.5$ kJ for each degree of freedom. Hence for argon it is 7.5 kJ mol^{-1} and for O$_2$ and N$_2$ it is 12.5 kJ mol^{-1}.

Example 3.6 Find the internal energy of a 1 kg mass of dry air at STP.
Answer: We can use $U = (f/2)\rho R_d TV = (f/2)\mathcal{M}R_d T_0$, where $\mathcal{M}$ is the mass of the gas in the system (here 1 kg) and f is 5. Then $U = 1.96 \times 10^5$ J $= 196$ kJ.   □

Example 3.7 Compare the rise in potential energy due to lifting the 1 kg parcel to 9 km, approximately one scale height.
Answer: The change in potential energy is $\mathcal{M}gh = 1.08 \times 10^5$ J $= 108$ kJ. Hence the gravitational potential change is comparable to the internal energy for a lift of one scale height.   □

3.3.2 Heat capacities

If a system composed of an ideal gas absorbs heat at a constant volume, its temperature will increase. Since the volume is held constant, the system can do no work on its environment in the process, therefore

$$\boxed{(\Delta U)_v = Q_{A \to B} = \mathcal{M}c_v \Delta T} \quad \text{[constant volume].} \qquad (3.13)$$

Differentiating the equation for the internal energy of an ideal gas (3.9) we obtain

$$\boxed{c_v = \frac{f}{2}R} \quad \text{[specific heat at constant volume and } R\text{].} \qquad (3.14)$$

Note that R as used in this equation is for a particular gas such as dry air. The heat capacity at constant volume, c_v, is proportional to the number of degrees of freedom.

Another important process is the heating of the gas at constant pressure. In this case

$$\boxed{Q_{A \to B} = \mathcal{M}c_p \Delta T} \quad \text{[heating at constant pressure].} \qquad (3.15)$$

We also have

$$W_{A \to B} = p\Delta V \qquad (3.16)$$

and

$$\Delta V = \mathcal{M}\frac{R\Delta T}{p}. \qquad (3.17)$$

This leads to

$$\Delta U = \mathcal{M}c_p \Delta T - \mathcal{M}R\Delta T. \qquad (3.18)$$

If U is a *function of state* as the First Law claims,[5] then it is given by $\mathcal{M}c_v \Delta T$, and we have an identity:

$$c_v \Delta T = c_p \Delta T - R\Delta T \qquad (3.19)$$

or

$$\boxed{R = c_p - c_v} \quad [\text{also } R^* = \bar{c}_p - \bar{c}_v] \qquad (3.20)$$

which is a very important relation for ideal gases, holding independently of the value of f. This last tells us that for an ideal gas

$$\boxed{c_p = \left(\frac{f}{2} + 1\right)R} \quad \left[\text{also } \bar{c}_p = \left(\frac{f}{2} + 1\right)R^*\right]. \qquad (3.21)$$

That c_p is always greater than c_v has an easy interpretation: some of the heat absorbed in the isobaric case is "wasted" by the expansion (work done by the system on the environment) rather than being devoted to raising the temperature.

Example 3.8 Dry air: what are c_v, c_p, using ideal gas rules?
Answer: We have $c_v = \frac{5}{2}R_d = 717.5\,\mathrm{J\,kg^{-1}\,K^{-1}}$, and $c_p = (f/2 + 1)R_d = c_v + R_d = (717 + 287) = 1004\,\mathrm{J\,kg^{-1}\,K^{-1}}$. □

Example 3.9 How much heat is required to raise the temperature of a 1 kg parcel of air at constant pressure (constant altitude in the atmosphere) by 1°C?
Answer: 1004 J. □

Example 3.10 A mass of 2 kg of dry air is heated isobarically from a temperature of 300 K to 310 K. How much heat is required?
Answer:

$$Q = \mathcal{M}c_p \Delta T = (2\,\mathrm{kg})(1004\,\mathrm{J\,kg^{-1}K^{-1}})(10\,\mathrm{K}) = 20080\,\mathrm{J} = 20.1\,\mathrm{kJ}. \qquad (3.22)$$

□

---

[5] A subtle thing is about to happen here: even though the process is occurring at constant pressure from one temperature to another, we can use the formula $\Delta U = \mathcal{M}c_v \Delta T$. This is because U is a function of state. No matter how we go from T_A to T_B, the change in U is the same. Here we see one of the most powerful tools in thermodynamics, the invoking of a state function's being a function only of its state (not path). Remember, however, that for many substances other than the ideal gas U might depend on more than just temperature; nevertheless, it is only a function of state.

Example 3.11 For Example 3.10, what is the change in internal energy?
Answer:

$$\Delta U = \frac{f}{2} M R_d \Delta T = \frac{5}{2} \times (2\,\text{kg})(287\,\text{J kg}^{-1}\text{K}^{-1})(10\,\text{K})$$

$$= 14350\,\text{J} = 14.35\,\text{kJ}. \tag{3.23}$$

☐

Example 3.12 For Example 3.10, what is the amount of work done?
Answer: Use the First Law:

$$W = Q - \Delta U = 5.72\,\text{kJ}. \tag{3.24}$$

☐

Example 3.13 A column of dry air 1 km thick (approximate thickness of the atmospheric boundary layer) and unit cross-section (1 m^2) is heated by sunlight at a rate $dQ/dt = 250\,\text{J s}^{-1}$. If the heating takes place at constant pressure ($p = 1000\,\text{hPa}$, $T = 300\,\text{K}$), by how many degrees per day is the column heated? (Assume the air above and the ground below are insulated from the well-mixed air in the boundary layer.)
Answer: $M = 1.16 \times 10^3$ kg,

$$\frac{dT}{dt} = \frac{dQ/dt}{c_p M}$$

where dQ/dt is the rate of heating (J s^{-1}). Then $dT/dt = 2.14 \times 10^{-4}$ K s$^{-1} = 18.5\,\text{K day}^{-1}$. This of course is a large rate of increase for normal conditions. The assumption of 250 W is the problem. The next example shows a more realistic case in which the heating is accomplished by black carbon particles in the air. Much of the heating in the atmospheric boundary layer comes from the absorption of infrared radiation as well as solar radiation by water vapor. ☐

Example 3.14: black carbon aerosol in air Suppose there are 100 black carbon particles per cubic centimeter in the air (10^8 particles m^{-3}). Let us take the radius of one of these particles to be 1 μm. This means the cross-sectional area of an individual particle is 3.14×10^{-12} m^2. The total area of intercepting carbon in a 1 m^3 block of air is 3×10^{-4} m^2. (Note that this is only a tiny fraction of the 1 m^2 cross-sectional area of the cube of air.) If the sun is straight overhead its flux is 1370 W m^{-2}. The heating rate of this cubic meter of aerosol-loaded air is then 0.43 W. If the air is at sea level its density is about 1.2 kg m^{-3}, and

$c_p = 1004 \, \text{J}\,\text{kg}^{-1}\,\text{K}^{-1}$. After some arithmetic we find that the air will experience an increase of temperature of $1.29 \, \text{K}\,\text{h}^{-1}$.   □

Calculus refresher: the natural logarithm The natural log of x is denoted by $y = \ln x$ (Figure 3.5) and is defined by

$$e^y = x.$$

We can deduce a few values, for example, for $x = 1, y = 0$, as $x \to \infty, y \to \infty$ and y is only defined for x positive. As $x \to 0, y \to -\infty$. Moreover,

$$\frac{dx}{dy} = e^y = x.$$

We can turn this into $dy = dx/x$, and then integrate:

$$\int_0^y dy = \int_1^x \frac{dx}{x}$$

which leads to an alternative definition of $\ln x$:

$$\ln x = \int_1^x \frac{dx}{x}.$$

Now a few properties. First the wonderful identity: $x = e^{\ln x}$ which follows from the definition. This can be used to derive a number of useful things: $ab = e^{\ln a}e^{\ln b} = e^{\ln a + \ln b} = e^{\ln ab} \Rightarrow \ln ab = \ln a + \ln b$. $x^a = e^{\ln x^a} = (e^{\ln x})^a = e^{a \ln x} \Rightarrow \ln x^a = a \ln x$. Finally, we already found that

$$\frac{d}{dx} \ln x = \frac{1}{x}.$$

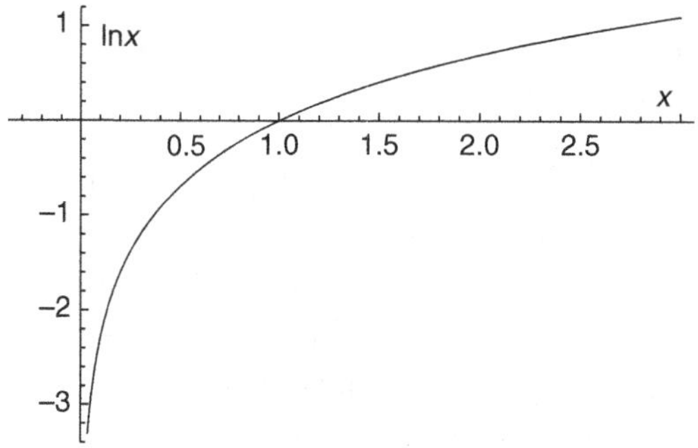

Figure 3.5 The natural logarithm $\ln x$ as a function of x.

3.3.3 Adiabatic processes and potential temperature

Many processes in meteorology involve parcels of air moving in such a way that no heat is exchanged with the environment during passage from one location to another, i.e., $Q = 0$. The parcel effectively is surrounded by a thermally insulating blanket. A key point here is that it takes a long time for temperature differences to diffuse into a parcel from outside compared to the relatively short time for the pressures inside and outside to equalize. This means that in many situations we can regard the process as being *adiabatic*, that is, isolated from diathermal contact. In that case for an infinitesimal displacement

$$dU = 0 - pdV \tag{3.25}$$

or

$$\mathcal{M}c_v\, dT = -pdV = -\frac{\mathcal{M}RT}{V}\, dV \tag{3.26}$$

or

$$\frac{dT}{T} = -\frac{R}{c_v}\frac{dV}{V}. \tag{3.27}$$

The next steps become smoother if we use the identity $c_v = c_p - R$, leading to:

$$(c_p - R)\frac{dT}{T} = -R\frac{dV}{V}. \tag{3.28}$$

Next we use the Ideal Gas Law

$$pV = \mathcal{M}RT \tag{3.29}$$

and take natural logs to obtain

$$\ln p + \ln V = \ln \mathcal{M}R + \ln T. \tag{3.30}$$

Taking the differentials we find:

$$\boxed{\frac{dp}{p} + \frac{dV}{V} = \frac{dT}{T}} \quad \text{[logarithmic derivative].} \tag{3.31}$$

The term $d(\ln \mathcal{M}R)$ disappears because it is constant. Now we can substitute in our original adiabatic equation to eliminate the V dependence:

$$(c_p - R)\frac{dT}{T} = -R\left(\frac{dT}{T} - \frac{dp}{p}\right) \tag{3.32}$$

and finally,

$$c_p \frac{dT}{T} = R \frac{dp}{p}. \tag{3.33}$$

After dividing through by c_p and defining

$$\boxed{\kappa = \frac{R}{c_p}} \quad [0.286 \text{ for dry air}] \tag{3.34}$$

we can integrate from (T_0, p_0) to (T, p) to find

$$\ln \frac{T}{T_0} = \kappa \ln \frac{p}{p_0} \tag{3.35}$$

or taking the anti-log: [6]

$$\boxed{T = T_0 \left(\frac{p}{p_0} \right)^{\kappa}} \quad [\text{Poisson's equation}]. \tag{3.36}$$

In atmospheric science the formula (3.36) is very important; one usually sees it in the form

$$\frac{T}{\theta} = \left(\frac{p}{1000 \text{ hPa}} \right)^{\kappa} \quad p \text{ in hPa} \tag{3.37}$$

where θ ($\theta = T_0$ at $p = 1000 \text{ hPa}$) is called the *potential temperature*. We often see it in the following form:

$$\boxed{\theta \equiv T \left(\frac{p_0}{p} \right)^{\kappa}} \quad [\text{potential temperature}]. \tag{3.38}$$

The last equation is called *Poisson's equation*. It gives the temperature that a parcel of dry air would have if it were brought adiabatically to a pressure of 1000 hPa. No matter where the parcel lies as a piece of the environment, its potential temperature is well defined. If the parcel moves adiabatically its potential temperature will not change. When we find a quantity describing the system (defined here as the parcel) which does not change as the parcel moves about (in this case adiabatically), we refer to it as a *conservative property*. In practice parcels do move adiabatically in convective motions to a good approximation. Moreover,

---

[6] Taking the anti-log means raising each side to be the power of e: $y = f(x) \Rightarrow e^y = e^{f(x)}$. Now using $x = e^{\ln x}$ means the anti-log of $\ln x$ is just x.

Table 3.1 *Important formulas for ideal gases in adiabatic processes*

Variables	Formula
T, p	$T/T_0 = (p/p_0)^\kappa$
θ, T, p	$\theta = T\,(p_0/p)^\kappa$
p, V	$pV^\gamma = p_0 V_0^\gamma$
T, V	$TV^{\gamma-1} = T_0 V_0^{\gamma-1}$
κ	$R/c_p = 0.286$ for dry air
γ	$\frac{c_p}{c_v} = 1.400$ for dry air

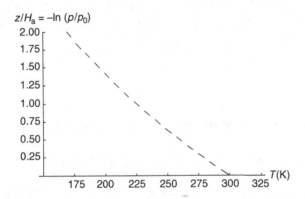

Figure 3.6 Dry adiabat for a parcel of air whose potential temperature is 300 K. The abscissa is T in K, and the ordinate is $-\ln(p/p_0) = z/H_a$. The value z/H_a is the height above sea level in units of the scale height, H_a, which is typically about 8 km in midlatitudes.

parcels moving horizontally move along constant θ surfaces (isentropic motion) (see Table 3.1).

Tip It might be difficult to remember the form of (3.38) (Poisson's equation), since it will appear in a variety of forms: is the κ with a negative sign? Is p_0 on top or not? etc. Just remember: let a parcel rise, in doing so θ stays constant, so that as T goes down, so must p. This can help us to obtain the correct formula.

Figure 3.6 shows how the temperature of a parcel is lowered if it is lifted in an atmosphere whose pressure follows $p(z) = p_0 e^{-z/H_a}$, which is a reasonable approximation to the actual behavior of the pressure. Here H_a is called the scale height of the atmosphere. Typically in midlatitudes, $H_a \approx 8$ km.

Example 3.15 Air in a jet plane near the tropopause is taken into the plane and compressed adiabatically from an outside pressure of 300 hPa to 1000 hPa. The temperature outside is 255 K. What must be done to the air to bring it to an inside temperature of 300 K?

Answer: In adiabatically compressing the gas the Kelvin temperature is raised by a factor of 1.411 to 360 K. The air must now be cooled at constant pressure to 300 K by the air conditioning system. Recalling that 3600 J = 1 Wh, this requires 60 kJ = 0.0167 kWh for each kilogram of air brought into the plane from the outside. ☐

Example 3.16 A 1 kg parcel of dry air is located at the 500 hPa level in the atmosphere. Its temperature is 246 K. What is its potential temperature?
Answer: $\theta = (1/2)^{-0.286}T$. We find $\theta = 300$ K. ☐

Example 3.17 A 1 kg mass of dry air is located at 500 hPa. Its potential temperature is $\theta = 300$ K (as above). 1 kJ of heat is absorbed by the parcel at constant pressure. What is the change in the parcel's potential temperature?
Answer: First, compute the change in temperature, $\Delta T = Q/(Mc_p) = 1$ kJ/$(1.004 \text{ kJ kg}^{-1} \text{ K}^{-1} \times 1 \text{ kg}) = 0.996$ K. Next, use Poisson's equation at 500 hPa:

$$\Delta\theta = \Delta T/(0.5)^{0.286} = 1.21 \text{ K}. \tag{3.39}$$

☐

Example 3.18 A 50 kg parcel of dry air has temperature 300 K at the surface (1000 hPa). It is lifted adiabatically to the 700 hPa level. What is its potential temperature?
Answer: $\theta = 300$ K. What is its temperature at the 700 hPa level?

$$T = \theta \left(\frac{p}{1000 \text{ mb}}\right)^{\kappa} = 300 \text{ K} \times (0.7)^{(0.286)} = 270.9 \text{ K}. \tag{3.40}$$

☐

Example 3.19 Air at 300 K is forced up a mountain slope adiabatically through a vertical height of 2 km. Suppose the pressure is given by $p(z) = p_0 e^{-z/H}, H = 10$ km. How much is the temperature changed? By what ratio is the volume of such a parcel changed?
   Answer: $p(2 \text{ km}) = p_0 e^{-2/10} = 0.819 p_0$. $T/300$ K $= (0.819)^{0.286}$. $T = 283$ K. $V_B/V_A = (T_B p_A)/(T_A p_B) = 1.15$. ☐

Example 3.20: other forms of the adiabatic curve Returning to the relation (3.27), use $R = c_p - c_v$, then divide through by c_v. We obtain

$$\frac{dT}{T} = -(\gamma - 1)\frac{dV}{V} \tag{3.41}$$

where $\gamma = c_p/c_v$ is called the ratio of specific heats. For an ideal diatomic gas (and air acts like one) $\gamma = 1.400$. We can integrate as in the text and obtain

$$\ln\frac{T}{T_0} = +(1 - \gamma)\ln\frac{V}{V_0} \tag{3.42}$$

and finally

$$\boxed{TV^{\gamma-1} = T_0V_0^{\gamma-1}} \quad \text{[adiabatic process } (T, V \text{ form)].} \qquad (3.43)$$

Another form useful in physics and engineering can be derived:

$$\boxed{pV^{\gamma} = p_0V_0^{\gamma}} \quad \text{[adiabatic process } (p, V \text{ form)].} \qquad (3.44)$$

$\square$

3.4 Enthalpy

In many meteorological and chemical applications the internal energy is not the most ideal state function for describing energetic changes during transitions. The classical form of the First Law is especially useful for transitions in which the volume is held fixed ($dV = 0$) since the volume-work term vanishes, but in atmospheric applications most changes in the state of a parcel occur either adiabatically or isobarically. Hence, it becomes convenient to introduce a new function of state called the *enthalpy*, H, defined by

$$\boxed{H \equiv U + pV} \quad \text{[enthalpy].} \qquad (3.45)$$

Take the differential to obtain

$$dH = dU + p\,dV + V\,dp. \qquad (3.46)$$

After substituting the earlier form of the First Law in terms of the internal energy we obtain

$$\boxed{dH = đQ + V\,dp} \quad \text{[enthalpy form of the First Law].} \qquad (3.47)$$

Very often atmospheric processes take place at a fixed pressure (altitude). These include heating of a parcel by solar radiation at a particular altitude, condensation heating, and contact heating at the surface. In this case the enthalpy is a very convenient function to describe the parcel's thermodynamic state. Note that the change in enthalpy under a constant pressure process is just

$$(dH)_p = đ_pQ = \mathcal{M}c_p dT \qquad (3.48)$$

or

$$\left(\frac{\partial H}{\partial T}\right)_p = \mathcal{M}c_p. \qquad (3.49)$$

Integrating the last equation:

$$H = \mathcal{M}c_p T + F(V) \tag{3.50}$$

where $F(V)$ is an arbitrary function of volume appearing here as an integration constant. We can find this function by noting that $U = \mathcal{M}c_v T$ and using the definition of H, (3.45), to obtain

$$\boxed{H = \mathcal{M}c_p T} \quad \text{[enthalpy for an ideal gas]}. \tag{3.51}$$

In other words, the arbitrary function $F(V)$ is identically zero for the ideal gas.
   The adiabatic process is expressed as

$$(\mathrm{d}H)_{\mathrm{d}Q=0} = 0 + V\,\mathrm{d}p \tag{3.52}$$

and for the ideal gas,

$$c_p\,\mathrm{d}T = \frac{RT}{p}\,\mathrm{d}p \tag{3.53}$$

and

$$c_p\frac{\mathrm{d}T}{T} = R\frac{\mathrm{d}p}{p} \tag{3.54}$$

which will quickly lead us to Poisson's equation (3.36).

Calculus refresher: partial derivatives Thermodynamic functions nearly always involve more than one variable as we have seen already, e.g., $V(T,p)$. The "partial" of $V(T,p)$ with respect to p holding T constant is defined by

$$\left(\frac{\partial V}{\partial p}\right)_T = \lim_{\Delta p \to 0} \frac{V(T, p+\Delta p) - V(T, p)}{\Delta p}.$$

In most fields the subscript T following the large parentheses is omitted, but in thermodynamics it is conventional (and useful) to retain this reminder of which variable is being held constant. Sometimes especially in mathematics and physics, the partial derivative is denoted by a subscript. For example, let f be a function of x and y, then $\partial f / \partial x = f_x$, etc. You simply take the ordinary derivative but hold all variables constant except the one being varied. For example, take the ideal gas: $V = \mathcal{M}RT/p$, then $(\partial V/\partial p)_T = -\mathcal{M}RT/p^2$.

This is a good time to search out your old calculus book and review the chapter on partial differentiation. An important result to remember is that if we go to second partial derivatives such as f_{xx} or f_{xy}, the *order* does not matter:

$$f_{xy} = f_{yx}. \tag{3.55}$$

The *differential* of $f(x, y)$ is

$$df = f_x \, dx + f_y \, dy. \tag{3.56}$$

If we divide through by dx and set dy to zero we obtain

$$\left(\frac{df}{dx} \right)_{dy=0} = f_x. \tag{3.57}$$

Suppose the function $f(x, y)$ is held constant. Then

$$f_x \, dx + f_y \, dy = 0 \tag{3.58}$$

and we find

$$\frac{dy}{dx} = -\frac{f_x}{f_y}. \tag{3.59}$$

The notation in the last equation will be encountered often.

Example 3.21 A 1 kg parcel is heated at the surface ($p = 1000$ hPa) at a rate $dQ/dt = 20 \, \text{W kg}^{-1}$ (W = watts). What is the rate of change of enthalpy?
Answer: Note that $dp/dt = 0$. Then, $dH/dt = dQ/dt$. ☐

Example 3.22 A parcel moves along an isobaric surface (constant pressure) and is heated at a rate $dQ_M/dt = 10 \, \text{W kg}^{-1}$. What is the rate of change of T along the path of motion?
Answer: $(dH/dt) = Mc_p dT/dt = dQ_M/dt$; $dT/dt = (dQ_M/dt)/c_p = 10 \, \text{W kg}^{-1}/1004 \, \text{J}^{-1} \, \text{kg}^{-1} = 0.01 \, \text{K s}^{-1}$. ☐

3.5 Standard enthalpy of fusion and vaporization

Enthalpy is a very useful function in describing the energy transfers in processes involving a change of phase (e.g., liquid to vapor). Enthalpy is especially useful since these processes often take place at constant pressure. An example is the evaporation of 1 mol of water. In this case the system (the volume containing the water) is heated by maintaining a small temperature differential with its surroundings at constant pressure and constant temperature. The heat (now we can call it enthalpy) absorbed to effect this transition is often called the *latent heat*

Table 3.2 *Standard enthalpies of transition for selected compounds*
The standard enthalpies of fusion and vaporization are evaluated at the freezing and boiling points (K) respectively at 1 atm of pressure. Units in this table for the enthalpies are $kJ\,mol^{-1}$, but be careful in using tables since the units of energy might be in kcal (4.18 kJ = 1 kcal). In the table T_f is the freezing point and T_b is the boiling point at 1 atm of pressure.

Species	T_f	$\Delta_{fus}\overline{H}^{\circ}$	T_b	$\Delta_{vap}\overline{H}^{\circ}$
CO_2	217.0	8.33	194.6	25.23 (sublimation)
H_2	3.5	0.021	4.22	0.084
H_2O	273.16	6.008	373.15	40.656
Ar	1.188	87.29	6.506	

of vaporization in the older literature, but in keeping with current convention it is called the *enthalpy of vaporization*. Many tables give values in terms of moles rather than kilograms of the substance. To make useful standardized tables, conventions have been adopted. In the case of evaporation for example, $\Delta_{vap}\overline{H}^{\circ}$ indicates that 1 mol of the substance is being considered (the overbar) and the superscript o indicates that it is at a standard temperature (units should be indicated in the table). See Table 3.2.

By definition, the *standard enthalpy for vaporization*, $\Delta_{vap}\overline{H}^{\circ}$, is the heat transferred to the system at constant pressure per mole in the process of vaporization of the substance from its liquid to its vapor form. The standard quantity is defined at the boiling point (373 K for water) at 1 atm of pressure.

Similarly the *standard enthalpy for fusion* is the heat transferred by the system to the surroundings at constant pressure per mole in the process of fusing the substance from its liquid to its solid form. It is labeled $\Delta_{fus}\overline{H}^{\circ}$. By convention the standard quantity is evaluated at the freezing point at 1 atm of pressure.

Example 3.23 We have 36 g of liquid water at 373 K and 1 atm of pressure. We wish to evaporate the water and raise its temperature to 473 K at constant pressure. What is the change in enthalpy for the two steps? ($c_{pvap} \approx 2\,kJ\,kg^{-1}\,K^{-1}$.)
Answer: 36 g is 2.0 mol of water. We must then give the system 2 mol × 40.656 $kJ\,mol^{-1}$ = 81.312 J for step 1. Step 2 is a heating at constant pressure. $\Delta H_2 = \mathcal{M}c_p\Delta T = 0.036\,kg \times 2\,kJ\,kg^{-1}\,K^{-1} \times 100\,K = 7.2\,kJ$. □

Example 3.24 Two grams of liquid water are evaporated into 1 kg of dry air at 1 atm constant pressure. The temperature is 300 K. What is the change in enthalpy of the system?

Answer: The dry air is irrelevant. We must do several steps to accomplish our goal. To use the value in Table 3.2 we must heat the 2 g of water to its boiling point. The change in enthalpy for this is $\Delta H_1 = \mathcal{M}c_{\text{liq}}\Delta T = 2\,\text{g} \times 4.18\,\text{J}\,\text{K}^{-1}\text{g}^{-1} \times 73\,\text{K} = 610\,\text{J}$. Next the water is evaporated: $\Delta H_2 = (2/18)\,\text{mol}\; 40.7\,\text{kJ}\,\text{mol}^{-1} = 4.52\,\text{kJ}$. In step 3 we must cool the vapor back down to its starting temperature: $\Delta H_3 = 2\,\text{kJ}\,\text{K}^{-1}\,\text{kg}^{-1} \times 0.002\,\text{kg} \times (-73\,\text{K}) = -292.0\,\text{J}$. Combining all three steps: $\Delta H_1 + \Delta H_2 + \Delta H_3 = 4.84\,\text{kJ}$. ☐

Example 3.25 In the last example, what is the temperature and density change for the original 1 kg of air which holds the 2 g of liquid water?
Answer: $\Delta T = \Delta H \mathcal{M}^{-1}c_p^{-1} = 4.84\,\text{kJ}/(1\,\text{kg}\;1004\,\text{J}\,\text{kg}^{-1}\text{K}^{-1}) = 4.5\,\text{K}$. The change in density is $\Delta\rho = -p\Delta T/R_d T^2 = 10^5\,\text{Pa} \times 4.5\text{K}/(287\,\text{J}\,\text{kg}^{-1}\text{K}^{-1})(300\,\text{K})^2 = 0.019\,\text{kg}\,\text{m}^{-3}$. This represents a 1.6% change in the density, enough to cause important buoyancy effects. ☐

In these examples we made liberal use of the fact that the enthalpy of a system depends only on its state, not on the path through which the state is found. This freedom allows us to use standard tables.

Notes
Most of the good thermodynamics books referred to in earlier chapters work well for this one.

Notation and abbreviations for Chapter 3

c_v, c_p	specific heats (heat capacity per kg) at constant volume, pressure $(\text{J}\,\text{kg}^{-1}\text{K}^{-1})$
$\bar{c}_v, \bar{c}_p$	molar specific heats $(\text{J}\,\text{mol}^{-1}\text{K}^{-1})$
C_v, C_p	heat capacities at constant volume, pressure $(\text{J}\,\text{K}^{-1})$
$(dH)_p$	change in enthalpy at constant pressure (J)
$dH/dt, dQ/dt$	time rate of change of enthalpy, heat transfer rate $(\text{J}\,\text{s}^{-1})$
đQ, đW	differentials for heat, work, the bar emphasizes path dependence (J)
ΔV	change in volume (m^3)
$\Delta_{\text{fus}}\bar{H}^\circ(X)$	standard enthalpy of fusion of substance X $(\text{J}\,\text{mol}^{-1})$
$\Delta_{\text{vap}}H$	change in enthalpy during vaporization (J)
$\Delta_{\text{vap}}\bar{H}^\circ(X)$	standard enthalpy of vaporization of substance X, the overbar indicates 1 mol of substance, the superscript o indicates at standard conditions (usually 25 °C) $(\text{J}\,\text{mol}^{-1})$
Δx	displacement in x
f	number of degrees of freedom of a molecule

f_x	partial derivative with respect to x
F	force (N)
γ	ratio of specific heats c_p/c_v (dimensionless)
H	enthalpy (J)
H_a	scale height of the atmosphere
k_B	Boltzmann's constant
$\kappa = R/c_p$	(dimensional)
κ_H	thermal conductivity (J m K^{-1})
$\mathcal{M}$	bulk mass (kg)
n_0	number density (molecules m^{-3})
ν	number of moles
p	pressure (Pa, hPa)
$dQ_{\mathcal{M}}/dt$	rate of heating per unit mass (J kg^{-1} mol^{-1} s^{-1})
R, R_G, R_d, R^*	gas constant (J kg^{-1} K^{-1}), gas constant for a gas G, for dry air, universal gas constant (J K^{-1} mol^{-1})
ρ	density (kg m^{-3})
T	temperature (K)
θ	potential temperature (K)
V	volume (m^3)
V_A, V_B	initial and final volumes
$W_{V_A \rightarrow V_B}$	work in going from V_A to V_B
W, Q	work done by the system, heat taken into the system

Problems

3.1 Suppose $p(z) = p_0 e^{-z/H}$. Evaluate the following.
   (a) dp/dz at $z = H/3$.
   (b) $\int_0^\infty p(z)\, dz$.
   (c) $\int_{H/2}^H p(z)\, dz$.
   (d) $\partial p/\partial H$ at $z = H/2$.
3.2 Let $p = \rho RT$. Evaluate the following.
   (a) $\partial p/\partial T$.
   (b) $\partial \rho/\partial T$.
   (c) $\partial p/\partial (1/T)$.
3.3 The *compressibility* of a substance is defined by

$$\kappa_X = -\frac{1}{V}\left(\frac{\partial V}{\partial p}\right)_X \tag{3.60}$$

where X is the variable being held constant. We can compress the gas isothermally (κ_T) or adiabatically (κ_θ) (θ is the potential temperature). Calculate both for an ideal gas.

3.4 The *coefficient of expansion* is defined by

$$\beta = \frac{1}{V}\left(\frac{\partial V}{\partial T}\right)_p. \tag{3.61}$$

Compute β for an ideal gas.

3.5 Show that for any gas

$$\left(\frac{\partial p}{\partial T}\right)_V = \frac{\beta}{\kappa_T}. \tag{3.62}$$

(*Hint:* $dV = \left(\frac{\partial V}{\partial p}\right)_T dp + \left(\frac{\partial V}{\partial T}\right)_p dT$, see the Calculus refresher in this chapter.)

3.6 Find the internal energy of 1 kg of dry air at STP.

3.7 Suppose the atmosphere has its pressure given by $p(z) = p_0 e^{-z/H}$ with $p_0 = 1$ atm and $T(z = 0) = 273$ K. Now suppose a 1 kg parcel is lifted adiabatically one scale height H. How much work does the parcel do on the environment in the process? What is the change in its specific internal energy, Δu?

3.8 *Isobaric process* A 1 kg parcel of dry air has temperature 285 K and pressure 1000 hPa. It is heated by contact with the dry ground to a temperature of 295 K. (a) What is Q? (b) What is the change of the parcel's specific internal energy? (c) What is the change in the parcel's specific enthalpy?

3.9 *Isothermal process* 1 kg of dry air at 300 K and 1000 hPa is expanded isothermally (pretty unusual in the atmosphere) from a volume of 2 m^3 to twice that value. (a) What is the work done by the gas in this expansion? (b) What is the heat absorbed? (c) What is the change in enthalpy?

3.10 *Isochoric process* 1 m^3 of dry air at 1000 hPa and 290 K is enclosed in a rigid box. What is its density (kg m^{-3})? 3 g of liquid water are evaporated into the box. What is the increase of temperature in this box whose volume is held fixed? What are the changes in internal energy and enthalpy? Sketch a diagram of the change in the V–p plane.

3.11 *Adiabatic process* A parcel of mass 1 kg is lifted adiabatically from 800 hPa, where its temperature is 270 K, to 600 hPa. What is the new temperature? What are the changes in internal energy and enthalpy? Sketch a diagram of the change in the V–p plane.

3.12 *Isobaric process* A parcel of essentially dry air is held at a fixed altitude where the pressure is 950 hPa. It is heated by infrared radiation being absorbed by some water vapor in the parcel. The heating rate is 20 J kg^{-1} s^{-1}. What is the rate of change of the temperature, specific enthalpy and specific internal energy? How does the potential temperature change per unit time? Sketch a diagram of the change in the V–p plane.

3.13 An air column is composed of dry air and the density of the air is given by $\rho(z) = \rho_0 e^{-z/H}$, where $\rho_0 = 1.25$ kg m^{-3} and $H = 10$ km.
(a) What is the mass of air (kg) lying above 1 m^2?
(b) How many idealized "air" molecules are above the 1 m^2?

(c) Then approximately how many "air molecules" are there in the entire Earth's atmosphere?

3.14 The speed of sound in air can be computed from the formula $v_{sound} = \sqrt{1/\rho\kappa_X}$, where κ_X is the compressibility holding the parameter X constant. Compare the sound speeds (taking $\rho = p/RT$) when $X = \kappa_T$ and $X = \kappa_\theta$. The latter fits the data. Do you recall from physics why the adiabatic compressibility gives the correct answer instead of the isothermal compressibility? See Problem 3.3 above.

3.15 Suppose the atmosphere satisfies $p(z) = p_0 e^{-z/H}$ and that it is isothermal ($T(z) = T_0$). What is the potential temperature θ as a function of z? Sketch a graph.

4

The Second Law of Thermodynamics

The Second Law of Thermodynamics deals with changes in the conditions of a system and its surroundings under transitions from one thermodynamic state to another. We will introduce a new function of state by considering a simple quasi-static series of changes for an ideal gas. First recall that some quantities are functions of the state only, while others depend on the path taken between two states. According to the First Law, the internal energy U is a function only of state with changes in going from state A to state B, denoted ΔU, depending only on the initial state coordinates (e.g., p_A, V_A, T_A) and the final state coordinates (p_B, V_B, T_B). Recall that the enthalpy H is also a function of state. Changes in other quantities such as the amount of work done by the system on the surroundings in the (quasi-static) transition depend upon the path taken between the two states. In this case the work done by the system on the surroundings may be written,

$$\mathcal{W}_{A \to B} = \int_A^B p(V) \mathrm{d}V \tag{4.1}$$

and the function $p(V)$ is a specific curve in the $V-p$ plane defining the path taken in going from A to B. The path must be quasi-static and reversible, otherwise the path $p(V)$ may not even be defined, nor could we use $p \mathrm{d}V$ in computing $\mathrm{d}\mathcal{W}$. The change in internal energy depends only on the difference

$$\Delta U = U_B - U_A \tag{4.2}$$

no matter what the path (even if the process is irreversible). For example, in the case of the ideal gas the internal energy (and enthalpy) depend only on the temperature, independent of the history of the system. For more complex systems U and H might depend on other state variables as well, but not on the history of the system. Note that the heat (induced by a temperature difference between the system and its environment) taken into the system during the transition must also depend on the

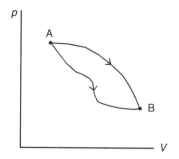

Figure 4.1 Illustration of two paths joining points A and B in a V–p state diagram.

path since

$$Q_{A \to B} = W_{A \to B} + U_B - U_A \qquad (4.3)$$

and the work done by the system surely depends on path as we have seen earlier (see Figure 4.1). Next we explore quasi-static transitions of an ideal gas to see whether there might be another function of the state of the system.

Consider the case of an *ideal gas* where the mass $\mathcal{M}$ is fixed. We can use the definition of the enthalpy to obtain:

$$dH = \mathcal{M}c_p \, dT = dQ + V dp. \qquad (4.4)$$

If we multiply through by the *integrating factor* $1/T$ we obtain

$$+\mathcal{M}c_p \frac{dT}{T} = \frac{dQ}{T} + V \frac{dp}{T}. \qquad (4.5)$$

Using the Ideal Gas Law, $V/T = \mathcal{M}R/p$ and solving for dQ/T:

$$\frac{dQ}{T} = -\mathcal{M}R\frac{dp}{p} + \mathcal{M}c_p\frac{dT}{T}$$

$$= -\mathcal{M}R\,d\ln p + \mathcal{M}c_p\,d\ln T$$

$$= \mathcal{M}c_p \left[d \ln \left(\frac{T}{p^{R/c_p}} \right) \right]$$

$$= d \left[\mathcal{M}c_p \ln(Tp^{-\kappa}) \right] \qquad (4.6)$$

where as before $\kappa = R/c_p$. The last expression says that dQ/T is a *perfect differential*. That is, its change in going from A to B does not depend on the path chosen for the sequence of quasi-static infinitesimal steps. A necessary and sufficient condition for the differential to be perfect is that integrals around arbitrary closed loops result in no change in the function.

Calculus refresher: perfect differential Suppose F is defined by:

$$F = \int_{\text{path}} A(x, y)\, dx + \int_{\text{path}} B(x, y)\, dy \qquad (4.7)$$

where $A(x, y)$ and $B(x, y)$ are well-behaved functions. It might happen that there exists a function $f(x, y)$ such that

$$\frac{\partial f}{\partial x} = A(x, y) \quad \text{and} \quad \frac{\partial f}{\partial y} = B(x, y), \qquad (4.8)$$

if so

$$df = \frac{\partial f}{\partial x}\, dx + \frac{\partial f}{\partial y}\, dy, \quad \text{or} \quad \frac{\partial A}{\partial y} = \frac{\partial B}{\partial x} \qquad (4.9)$$

is the exact differential of f and

$$F = \int_{\text{path}} df = f(\text{upper}) - f(\text{lower}) \qquad (4.10)$$

which means that F is independent of the path along which the integral is taken joining the upper and lower limits. It might be easier to see this if we introduce a parameter t which defines the curve

$$x = x(t), \qquad y = y(t) \qquad (4.11)$$

(for example, a parabola $y = x^2$ could be written $x(t) = t, y(t) = t^2$) this defines a curve in the x–y plane (the path). The integral can be written:

$$
\begin{aligned}
F &= \int_{t_1}^{t_2} \left(A\frac{dx}{dt} + B\frac{dy}{dt} \right) dt \\
&= \int_{t_1}^{t_2} \left(\frac{\partial f}{\partial x}\frac{dx}{dt} + \frac{\partial f}{\partial y}\frac{dy}{dt} \right) dt \\
&= \int_{t_1}^{t_2} \frac{df}{dt}\, dt = \int_{f(x(t_1),y(t_1))}^{f(x(t_2),y(t_2))} df \\
&= f(x(t_2), y(t_2)) - f(x(t_1), y(t_1)). \qquad (4.12)
\end{aligned}
$$

Example 4.1 Consider the function

$$f(x, y) = \sin xy. \qquad (4.13)$$

Then $f_x(x, y) = y \cos xy$ and $f_y(x, y) = x \cos xy$. Then

$$df = y \cos xy \, dx + x \cos xy \, dy = d(\sin xy). \tag{4.14}$$

□

Example 4.2: integrating factor Consider the first-order linear differential equation for the function $y(t)$:

$$\frac{dy}{dt} + by = g(t) \tag{4.15}$$

where b is a constant and $g(t)$ is a given function. Now multiply through by the *integrating factor* e^{bt}. We can show that

$$\frac{d}{dt}\left\{e^{bt}y\right\} = e^{bt}g(t) \tag{4.16}$$

and the left-hand side is rendered a perfect differential, by virtue of our knowing the proper integrating factor. Now we are in position to solve the differential equation by integrating each side:

$$e^{bt}y(t)\Big|_{t_0}^{t} = \int_{t_0}^{t} e^{bt'}g(t') \, dt'. \tag{4.17}$$

Then

$$y(t) = y(t_0)e^{-b(t-t_0)} + e^{-bt}\int_{t_0}^{t} e^{bt'}g(t') \, dt'. \tag{4.18}$$

□

Example 4.3 Consider the differential expression

$$dZ = 2xy^3 \, dx + 3x^2y^2 \, dy. \tag{4.19}$$

We want to consider integrals of this differential from the point (0,0) to the point (1,1) along some different paths. First consider the straight line path, $y = x$, which connects the two points. Then,

$$dZ = (2x^4 + 3x^4) \, dx = 5x^4 \, dx, \tag{4.20}$$

$$\Delta Z_{\text{straight line}} = \int_{0}^{1} 5x^4 \, dx = 1. \tag{4.21}$$

Next consider the parabolic path, $y = x^2, dy = 2x \, dx$, which also passes through the two points:

$$dZ = 2x^7 \, dx + 3y^3 \, dy, \tag{4.22}$$

$$\int_{parabolic} dZ = \int_0^1 2x^7 \, dx + \int_0^1 3y^3 \, dy = \frac{2}{8} + \frac{3}{4} = 1. \qquad (4.23)$$

We obtain the same answer for the two different paths. In fact, it is possible to write

$$dZ = d(x^2 y^3). \qquad (4.24)$$

In other words, $dZ(x, y)$ is a perfect differential. Integrating from one point in the x–y plane to another always yields the same answer.

On the other hand, suppose we had the function

$$dP = 2xy \, dx + 3x^2 \, dy. \qquad (4.25)$$

This differential is not perfect, but with the aid of the integrating factor, y^2, we can create the perfect differential, dZ.   □

Example 4.4 In classical mechanics we encounter the same concept with *conservative forces*, which implies the existence of a potential energy function. If a force can be described by a potential energy function $V(x, y, z)$ and if the force, $F(x, y, z)$ can be written as the gradient of the force:

$$F(x, y, z) \cdot dr = -\nabla V(x, y, z) \cdot dr$$
$$= -\left(\frac{\partial V}{\partial x}\right) dx - \left(\frac{\partial V}{\partial y}\right) dy - \left(\frac{\partial V}{\partial z}\right) dz$$
$$= -dV. \qquad (4.26)$$

Or we can write

$$F(r) = -\nabla V(r). \qquad (4.27)$$

Hence the work done in going from A to B:

$$W_{A \to B} = \underbrace{\int_A^B F \cdot dr}_{path}$$
$$= \underbrace{V(A) - V(B)}_{\text{independent of path}}. \qquad (4.28)$$
□

4.1 Entropy

In the case of the ideal gas mentioned above, we can use Poisson's equation,

$$T = \theta \left(\frac{p}{p_0}\right)^\kappa \qquad (4.29)$$

or

$$Tp^{-\kappa} = \theta (p_0)^{-\kappa} \tag{4.30}$$

and inserting into (4.6) we obtain:

$$\frac{dQ}{T} = \mathcal{M}c_p \, d \ln \left(\theta (1000 \, \text{hPa})^{-\kappa} \right)$$

$$= \mathcal{M}c_p \, d \ln \theta$$

$$\equiv dS \tag{4.31}$$

where S is called the *entropy*. In chemistry and physics the entropy is usually denoted S and the entropy per unit mass s; in some older meteorology contexts it is denoted ϕ. The important thing is that entropy is a function only of state and that its change in going from one state to another can be calculated by choosing a reversible path joining initial and final states.

For an infinitesimal displacement along a reversible path (for a general thermodynamic system, not just an ideal gas),

$$\boxed{dS = \frac{dQ_{\text{rev}}}{T}} \quad \text{[Second Law of Thermodynamics]}. \tag{4.32}$$

The extra subscript rev is added to remind the reader that the calculation of dQ/T must be along a reversible path. Recall that a reversible path is one in which the infinitesimal steps along the path are quasi-static and such that each can be reversed to restore the system to its previous state. In performing the calculation, we should find such an imaginary but realizable path joining the two states for the purposes of calculation. Of course, in nature the transition might be (and often is) an irreversible or spontaneous one from state A to state B, but there always exists a reversible path joining the two states so that the change in entropy can always be calculated. As indicated in the last equation, the very existence of such a function of state as defined above constitutes one statement of the Second Law of Thermodynamics.

In the case of dry parcels of ideal gas (nearly always satisfied in the dry atmosphere) we can use the formula

$$\boxed{dS = \mathcal{M}c_p d \ln \theta} \tag{4.33}$$

or

$$\boxed{S = \mathcal{M}c_p \ln \theta} \tag{4.34}$$

and in terms of specific quantities (kg^{-1}):

$$ds = c_p \, d \ln \theta, \qquad s = c_p \ln \theta, \tag{4.35}$$

where an arbitrary integration constant has been set to zero, since it is never actually needed. In these last formulas s refers to the entropy per unit mass (units $J\,K^{-1}\,kg^{-1}$) or the *specific entropy*. We will use $\bar{s}$ (units $J\,K^{-1}\,mol^{-1}$) to indicate entropy per mole. For a parcel of dry air, knowing the potential temperature is equivalent to knowing its entropy. Hence, the change in entropy for a parcel undergoing a reversible transition (through a series of quasi-static changes) can be computed simply by computing the change in its potential temperature

$$\Delta S = S_B - S_A$$

$$= Mc_p \ln \frac{\theta_B}{\theta_A}. \tag{4.36}$$

It is obvious that for a reversible adiabatic process, the change in entropy vanishes, since $\bar{d}Q = 0$ along the reversible path. For so-called *diabatic* processes (ones in which some heat is exchanged between the system and its surroundings), the formula defining the change in entropy can be used ($dS = dQ_{rev}/T$).

Example 4.5 Show that $dQ_{rev}/T = Mc_p\, d\ln\theta$ really works for an ideal gas. First write

$$dQ = dH - V\,dp. \tag{4.37}$$

Substituting for dH and V:

$$dQ = Mc_p dT - MRT \frac{dp}{p}. \tag{4.38}$$

After dividing through by T we simply repeat the steps leading to (4.6). ☐

Example 4.6 Compute the change in entropy for a parcel of mass M at pressure level p_0 being heated diabatically (e.g., by radiation heating) from temperature T_0 to T.
Answer: Since the heating is done at a fixed level, the pressure p remains constant at a value p_0. We can proceed to compute the change in entropy by using

$$dS = \frac{dQ}{T} \tag{4.39}$$

$$\Delta S = \int_{T_0}^{T} c_p M \frac{dT}{T}$$

$$= c_p M \ln \frac{T}{T_0}. \tag{4.40}$$
☐

Example 4.7 How much does the potential temperature of the parcel change in this heating if it takes place at 500 hPa?

Answer: Since $\theta = T (p/p_0)^{-\kappa}$, $p = $ constant $= 500$ hPa, and $p_0 = 1000$ hPa, we have $\Delta\theta = \Delta T (p/p_0)^{-\kappa} = (T - T_0) (500/1000)^{-0.287}$, for an isobaric heating process.   □

Example 4.8 A parcel of ideal gas is taken from (V_0, p_0) to $(2V_0, \frac{1}{2}p_0)$ by first (path a) expanding isobarically to $2V_0$, then the pressure is reduced at constant volume to $\frac{1}{2}p_0$ (path b). What is the change in entropy?

Answer: Along path a we compute $\Delta S_a = \int dQ/T = Mc_p \int_{T_0}^{2T_0} dT/T = Mc_p \ln 2$. Along path b, $\Delta S_b = \int dQ/T = \int dU/T = Mc_v \int_{2T_0}^{T_0} dT/T = -Mc_v \ln 2$. Now $\Delta S_a + \Delta S_b = M(c_p - c_v) \ln 2 = MR \ln 2$.   □

Example 4.9 Using the same initial and final states as in the previous example, compute the change in entropy but along the isothermal path joining the two states (path c).

Answer: Along c we have $\int dQ/T = \int dW/T = (1/T_0) \int_{V_0}^{2V_0} p \, dV = (MRT_0/T_0) \int_{V_0}^{2V_0} dV/V = MR \ln 2$, which is the same as in the alternative reversible two-step path chosen in the previous example.   □

Example 4.10: change in phase Compute the change in entropy for 1 kg of water being evaporated to gas at 373 K at constant pressure.

Answer: First note that 1 kg of water is 55.56 mol. The change in entropy is $Q/(373 \text{ K})$. Since the pressure is held constant, $Q = \nu \Delta_{\text{vap}} \overline{H}^\circ = 55.56 \text{ mol} \times 40.66 \text{ kJ mol}^{-1} = 2259 \text{ kJ}$. The change in entropy is then $\Delta S = 2259 \text{ kJ}/(373 \text{ K}) = 6.06 \text{ kJ K}^{-1}$.   □

4.2 The Second Law of Thermodynamics

There are several equivalent statements of the Second Law of Thermodynamics. For our present purposes it suffices to state the law as follows.

There exists a function S of the extensive parameters of the system that is a function of state and whose changes from one state to another can be calculated by

$$dS = \frac{dQ_{\text{rev}}}{T} \qquad (4.41)$$

and such that

$$dS \geq 0 \qquad (4.42)$$

for an isolated system.

The subscript is again a reminder that the calculation must be conducted along a reversible path.

4.3 Systems and reversibility

It is time to pause and review some definitions. Consider a *system* which is embedded in its *surroundings*. Together we say these comprise the *universe*. The system is in contact with its surroundings by various movable or stationary walls and membranes which might (or might not) allow fluxes (heat or mass) to cross. If no mass crosses we say the system is closed, whereas if mass crosses into or out of the volume confining the system we say it is an open system (sometimes this is called a *control volume*). So far in this book we have only considered closed systems, but we will encounter open systems as well. In equilibrium, both the system and the surroundings have thermodynamic coordinates. A reversible change is one which is quasi-static (taken in small slow steps in such a way that equilibrium is maintained; that is, there is enough time between steps for the pressure and temperature to homogenize throughout the volume of the system) and which can be reversed at any point returning both system and surroundings to their former coordinate values without the expenditure of any additional work (more on this below).

Often a system in contact with surroundings undergoes a spontaneous transition when a constraint is released or relaxed to some new configuration. Such a transition is irreversible. Under many of these spontaneous transitions the internal energy does not change, but the entropy does.

Example 4.11: Free expansion Consider a chamber isolated from the outside by adiabatic walls. Inside the chamber is a wall separating half the volume on each side. There is an ideal gas on one side of the partition, vacuum on the other. Suddenly the partition is removed (slipped out sideways so that no work is done), such that the gas expands (irreversibly) to fill the whole volume. What are the changes in internal energy, enthalpy and entropy?

Answer: The internal energy may be calculated from the First Law. The system does no work since the vacuum exerts no back pressure during the expansion. Also no heat is taken into the system because the walls are impermeable to such a transfer. Therefore, the internal energy does not change: $\Delta U = Q_{\text{free exp.}} - W_{\text{free exp.}}$. Note that we were able to apply the First Law even though the path was irreversible. The fact that the internal energy is invariant means that in the free expansion, the temperature does not change (true for an ideal gas – in a real gas there is some temperature change, even though no change in U occurs). Since the temperature does not change we can see that the enthalpy does not change for the ideal gas. The change in entropy must be computed by an alternative reversible path. There are many, but we can choose the one along an isotherm from V_0 to $2V_0$:

$$\Delta S = \underbrace{\int_{V_0}^{2V_0} \frac{dQ}{T}}_{\text{along isotherm}}. \tag{4.43}$$

Along the isotherm, $dQ = dW = p\, dV = \mathcal{M}RT\, dV/V$, therefore

$$\Delta S = \int_{V_0}^{2V_0} \mathcal{M}R\frac{dV}{V} = \mathcal{M}R\ln 2 > 0. \tag{4.44}$$

We could also compute the change in entropy for the surroundings (the adiabatic walls). It is zero. Hence, the change in entropy for the universe is $\mathcal{M}R\ln 2$ which is a positive number. $\qquad\qquad\qquad\qquad\qquad\qquad\qquad\qquad\qquad\qquad\qquad\Box$

In the free expansion process in the example above, the entropy of the system experienced a net increase, while that of the surroundings did not change. This is an example of the second part of the Second Law. *For an irreversible process the entropy of the system and its surroundings (taken together, the universe) will always increase.* Note that for an infinitesimal increase in volume that is an adiabatic free expansion, the entropy of the universe increases. This means that some quasi-static processes are irreversible. An irreversible process could be defined as one in which the entropy of the universe increases. So how could an adiabatic expansion ever be reversible? The way out is that one can imagine making infinitesimal (and reversible!) expansions at constant volume then at constant pressure in a stair-step procedure to approximate the adiabatic expansion curve in the V–p plane, much as an integral can be approximated by summing rectangular boxes whose upper edges approximate the curve being integrated. In this way a reversible approximation can be found to the adiabatic expansion curve.

4.4 Additivity of entropy

The entropy for a set of thermodynamic systems is the sum of the individual entropies of the constituent subsystems. Hence, the entropies of the system and its surroundings are *additive*. The same *additivity principle* applies to the internal energy and the enthalpy and any other of the extensive parameters describing the composite system. The extensive parameters volume and mass also satisfy this principle. When we add up several systems to form a larger system in which the entropies, internal energies and enthalpies can be added up, we call this a *composite system*. We can express this in equation form:

$$S = S_1 + S_2 + \cdots \tag{4.45}$$

$$U = U_1 + U_2 + \cdots \tag{4.46}$$

$$H = H_1 + H_2 + \cdots. \tag{4.47}$$

Example 4.12: a column of air We can think of a column of air as a composite system. Its pressure and temperature might be varying with altitude z but we can add up these contributions from individual slabs:

$$U_{\text{column}} = \int_{z_{\text{lower}}}^{z_{\text{upper}}} \rho(z)u(z)\,dz \tag{4.48}$$

$$H_{\text{column}} = \int_{z_{\text{lower}}}^{z_{\text{upper}}} \rho(z)h(z)\,dz \tag{4.49}$$

$$S_{\text{column}} = \int_{z_{\text{lower}}}^{z_{\text{upper}}} \rho(z)s(z)\,dz \tag{4.50}$$

where $\rho(z)$ is mass density, $u(z)$ is the specific internal energy, etc. □

In general, if two systems are brought into "contact" we can say the sum of the internal energies of the system will remain the same. Bringing two systems into contact amounts to removing or relaxing a constraint. For example, if two gases A and B at the same pressure are in a chamber, but separated by a partition, the removal of the partition will allow the gases to mix. No change in internal energy will occur if the chamber containing both subsystems is insulated from its environment. On the other hand, the change in the total entropy must be zero or positive. Processes (either reversible or not) in which the sum of the entropies of a system and its surroundings yield a negative change in total entropy do not occur in nature. This principle can be of great utility in determining which way a process will proceed when a constraint is relaxed or removed. Recall that removal of a constraint means that in order to restore the system to its original state additional work must be performed by some external agent. This last is really the essence of the Second Law of Thermodynamics. Solutions to problems of this type may not always be facilitated by use of the internal energy alone, since it may remain fixed when the constraint is removed. But restoration of the original conditions does require work, and hence changes in the internal energy alone are insufficient to describe what has happened. It turns out that the entropy change provides this additional information.

Example 4.13 How could we restore the gas in Example 4.11 to its original condition and how much work would be required?
Answer: First we would have to bring the ideal gas into contact with a reservoir of temperature T, then we would perform an isothermal compression of the gas from volume $2V_0$ to V_0. As we perform the compression of the gas, infinitesimal temperature differences will develop between the system and the reservoir: heat will transfer from one system to the other to maintain the fixed temperature. The work performed (by the system) in the isothermal compression is just $-\mathcal{M}RT \ln 2$.

Note that in the case of mixing two gases A and B mentioned above we would need to recompress each to its original volume by an isothermal route in order to restore them to their original states. We might in this case use a membrane which allows molecules A to pass into a new adjacent chamber but not B. But suppose A and B are the same gas. Does the entropy increase when the partition is removed? No.   □

4.4.1 Expression for entropy of an ideal gas

It is possible to derive an analytical expression for the entropy of an ideal gas. To proceed consider the expression for an ideal gas dry air parcel. Recall that the entropy S is given by

$$S = \mathcal{M}c_p \ln \frac{\theta}{\theta_0}. \tag{4.51}$$

Using the expression defining the potential temperature we have

$$S = \mathcal{M}c_p \ln \left[\frac{T}{T_0} \left(\frac{p}{p_0} \right)^{-\kappa} \right]. \tag{4.52}$$

While this is a perfectly good expression for the entropy, it is not in terms of extensive variables $\mathcal{M}, U, V$. Making use of the defining equations for an ideal gas: $p = \mathcal{M}RT/V$, $U = (f/2)\mathcal{M}RT$ and the ideal gas property $c_p = (f/2+1)R$, we eventually find

$$\boxed{S = \mathcal{M}R \ln \left[\left(\frac{U}{U_0} \right)^{f/2} \left(\frac{V}{V_0} \right) \right]} \quad \text{[entropy of an ideal gas]} \tag{4.53}$$

which holds for any ideal gas. For dry air, $R = R_d, f = 5$. To express this in molar form replace $\mathcal{M}R$ with νR^*. We immediately see that such a function S does exist for the simple ideal gas system. We also see that S is extensive (proportional to $\mathcal{M}$), and the extensive arguments U and V are expressed as ratios of standard states with the same mass. In addition we can see that $S(U, V)$ is an increasing function of the internal energy U and the volume V.

A more direct way of deriving the entropy of an ideal gas is to start with the First Law (now with đQ replaced with T dS; hence the expression that follows will hold for reversible paths in the V–S plane):

$$\boxed{dU = T\,dS - p\,dV} \quad \text{[differential for } U \text{ in terms of entropy].} \tag{4.54}$$

Dividing through by T and rearranging:

$$dS = \frac{1}{T}dU + \frac{p}{T}dV. \tag{4.55}$$

Using $U = (f/2)\mathcal{M}RT$ and $p = (\mathcal{M}RT/V)$:

$$dS = \frac{f}{2}\mathcal{M}R\frac{dU}{U} + \mathcal{M}R\frac{dV}{V}. \tag{4.56}$$

Then,

$$S = S_0 + \frac{f}{2}\mathcal{M}R\ln\left(\frac{U}{U_0}\right) + \mathcal{M}R\ln\left(\frac{V}{V_0}\right). \tag{4.57}$$

The last equation is equivalent to the expression (4.53) with $S_0 = 0$.

4.4.2 *Expression for the internal energy of an ideal gas*

We can solve for U using the last equation:

$$\boxed{\frac{U}{U_0} = \left(\frac{V}{V_0}\right)^{-2/f} \exp\left[\frac{S - S_0}{(f/2)\mathcal{M}R}\right]} \quad [U \text{ for an ideal gas}]. \tag{4.58}$$

This expresses $U = U(S, V, \mathcal{M})$ for a single-component system. The *thermo-dynamic potentials* are the intensive parameters of the system that come from taking partial derivatives of U with respect to its arguments:

$$\boxed{\left(\frac{\partial U}{\partial S}\right)_{V,\mathcal{M}} = T} \quad [\text{positive definite}] \tag{4.59}$$

$$\boxed{\left(\frac{\partial U}{\partial V}\right)_{S,\mathcal{M}} = -p} \quad [\text{negative definite}]. \tag{4.60}$$

The first leads to $T = U/((f/2)\mathcal{M}R)$ which defines U for the ideal gas. The second leads to the ideal gas equation of state, $pV = \mathcal{M}RT$.

While we worked out the analytical expressions for the case of an ideal gas, the differential relations for dU and dS are general and hold for any substance. Therefore the last two equations for the partial derivatives also hold generally. We then learn two important facts. (1) The internal energy is always an increasing function of S (likewise, S is an increasing function of U). (2) For constant entropy the internal energy is a decreasing function of V.

4.4.3 Enthalpy of an ideal gas

By a series of steps similar to those of the last subsection we can find:

$$H(S,p) = H_0 \left(\frac{p}{p_0}\right)^\kappa \exp\left(\frac{S - S_0}{\mathcal{M}c_p}\right) \quad \text{[enthalpy of an ideal gas].} \quad (4.61)$$

This expression shows explicitly that for a simple ideal gas, the enthalpy is a function of the entropy and the pressure.

In the general case (not just the ideal gas) we have for a single-component system

$$dH = T\,dS + V\,dp \quad \text{[differential of H in terms of entropy]} \quad (4.62)$$

and therefore

$$\left(\frac{\partial H}{\partial S}\right)_p = T, \quad \left(\frac{\partial H}{\partial p}\right)_S = V \quad (4.63)$$

An important corollary of the expressions for dU, dS and dH is that they are natural functions of certain pairs of variables (fixed mass). For example, $U = U(S,V), S = S(U,V)$ and $H = H(S,p)$.

4.5 Extremum principle

Since the entropies of the system and its surroundings together always increase in a natural (spontaneous) process, we can see that the final state after constraints have been released will be one in which the entropy is a maximum. For example, the equilibrium concentration of gaseous reactants and gaseous products in a chemical reaction will reach its equilibrium value when the entropy of the system is maximized (see Chapter 5).

Example 4.14 Consider two isolated parcels of dry air with masses $\mathcal{M}_1$ and $\mathcal{M}_2$ at the same pressure $p_1 = p_2$ (same altitude) but at different initial temperatures T_1^0 and T_2^0. The two subsystems come into thermal contact. What is the final temperature after this irreversible process takes place?

Answer: The enthalpy is constant during this mixing process

$$H = c_p \mathcal{M}_1 T_1^0 + c_p \mathcal{M}_2 T_2^0$$
$$= c_p(\mathcal{M}_1 + \mathcal{M}_2)T_F \quad (4.64)$$

where T_F is the resulting temperature of the mixture. We can solve for T_F:

$$T_F = \frac{\mathcal{M}_1 T_1^0 + \mathcal{M}_2 T_2^0}{\mathcal{M}_1 + \mathcal{M}_2} = xT_1^0 + (1 - x)T_2^0 \quad (4.65)$$

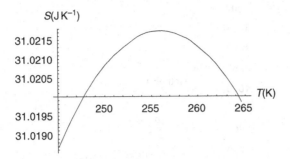

Figure 4.2 Entropy versus final temperature for component one of a mixture subject to the constraint that enthalpy be conserved.

where $x = M_1/M$, $M = M_1 + M_2$. The last equation is the answer to our problem, but let us look further to see what happens to the entropy. The potential temperature of subsystem i is $\theta_i = (p_0/p_i)^\kappa \times T_i$; $p_0 = 1000$ hPa. The entropy of the system is the sum of the entropies of the individual subsystems comprising the whole system

$$S = c_p M_1 \ln \theta_1 + c_p M_2 \ln \theta_2$$
$$= c_p \ln \left[(\theta_1)^{M_1} (\theta_2)^{M_2} \right]$$
$$= c_p \ln \left[(T_1)^{M_1} (T_2)^{M_2} \right] + \text{constant}$$
$$= c_p \ln \left[(T_1)^{Mx} (T_2)^{M(1-x)} \right] + \text{constant}. \tag{4.66}$$

$\square$

Consider a particular example for which $M_1 = 2\,\text{kg}$, $M_2 = 3\,\text{kg}$, $x = 0.4$, $T_1^0 = 250\,\text{K}$, $T_2^0 = 260\,\text{K}$ and for simplicity $c_p = 1$. The resulting $T_F = 256\,\text{K}$. Figure 4.2 shows a plot of S as a function of T_1. Our assumption based on intuition and experience that both masses come to the same temperature of 256 K is justified by its being the value of temperature that maximizes the entropy of the combined system.

4.5.1 *Carnot cycle*

The Carnot cycle is the most important closed loop process in thermodynamics. We illustrate it here for an ideal gas. This is a four step loop process as illustrated in Figure 4.3. The branch ab is along a hot isotherm at temperature T_h during which heat Q_h is transferred to the system from the hot reservoir. The next step is an adiabatic expansion bc to the cooler temperature T_c. The third step, cd is isothermal and an amount of heat $Q_c(> 0)$ is expelled from the system to the cooler reservoir at T_c. Finally, there is the adiabatic compression da, which completes the cycle

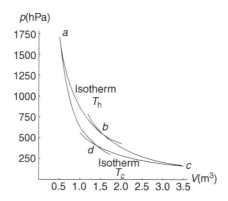

Figure 4.3 Carnot cycle for 1 kg of dry air (taken as an ideal gas) in the V–p plane. The cycle proceeds as follows. Step ab is an isothermal expansion from a to b, at temperature T_h, drawing in heat Q_h. Step bc is an adiabatic expansion from b to c. Step cd is an isothermal compression at temperature T_c expelling heat Q_c from the system. Finally, step da is an adiabatic compression from d to a. In this case $V_a = 0.50\,\mathrm{m^3}$, $T_2 = 300\,\mathrm{K}$, $V_b = 1.50\,\mathrm{m^3}$, $T_1 = 200\,\mathrm{K}$. Then all other intersections are determined, e.g., $p_a = 1720\,\mathrm{hPa}$, $p_b = 574\,\mathrm{hPa}$.

back to the starting point a. We can list the products:

$$\mathcal{W}_{ab}(>0),\ \mathcal{W}_{ab} = \mathcal{M}RT_h \ln \frac{V_b}{V_a} = Q_h(>0) \tag{4.67}$$

$$\mathcal{W}_{bc}(>0),\ \mathcal{W}_{bc} = -\Delta_{bc}U = \mathcal{M}c_v(T_h - T_c) \tag{4.68}$$

$$\mathcal{W}_{cd}(<0),\ \mathcal{W}_{cd} = \mathcal{M}RT_c \ln \frac{V_d}{V_c} = -Q_c(<0) \tag{4.69}$$

$$\mathcal{W}_{da}(<0),\ \mathcal{W}_{da} = -\Delta_{da}U = \mathcal{M}c_v(T_c - T_h) \tag{4.70}$$

$$\Delta_{ab}S = \frac{Q_h}{T_h}, \qquad \Delta_{cd}S = -\frac{Q_c}{T_c} \tag{4.71}$$

and since it is a closed loop, $\oint \mathrm{d}S = 0$, we have

$$\Delta_{ab}S + \Delta_{cd}S = 0 \tag{4.72}$$

which leads to

$$\boxed{\frac{T_h}{T_c} = \frac{Q_h}{Q_c}} \tag{4.73}$$

and in addition we can find from the above formulas:

$$\boxed{\frac{V_b}{V_a} = \frac{V_c}{V_d}}. \tag{4.74}$$

In the Carnot cycle we follow convention and call the heat transferred to the system along the hot isotherm ($a \rightarrow b$) Q_h and the heat rejected along the cooler isotherm $-Q_c$. This way both Q_h and Q_c are positive numbers. Note that $Q_h > Q_c$: more heat is always drawn from the hot reservoir than is rejected to the cold one. The difference is the work extracted from the system during the cycle. It would be nice to use all of the heat from the hot reservoir to create work, but this is impossible, since some heat is always rejected to the cold reservoir. We can never turn all of our heat extracted from the hot reservoir into work. This is a major consequence of the Second Law. The First Law says energy must be conserved, but the Second Law goes further to tell us that we not only cannot obtain something for nothing, we cannot even obtain all of the something to be used for work.

The *efficiency* of the process is the work performed over the whole loop divided by heat extracted from the hot reservoir:

$$\boxed{\text{efficiency} = \frac{Q_h - Q_c}{Q_h} = 1 - \frac{T_c}{T_h}} \tag{4.75}$$

Figure 4.4 shows (not to scale) the same cycle in the entropy–temperature plane, referred to as the S–T plane. In this plane the cycle is a more simple geometric figure, a rectangle. Consider the area enclosed in a closed figure in the S–T plane

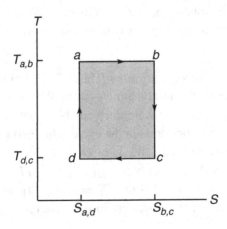

Figure 4.4 Schematic diagram of the Carnot cycle of Figure 4.3 except in the S–T plane. The geometrical figure is a rectangle independent of the substance (i.e., this diagram is not restricted to the ideal gas). The area enclosed is the work done by the system.

(such as Figure 4.4):

$$\boxed{\text{area enclosed} = \oint T(S)\, dS} \quad \text{[area-energy in the } S\text{--}T \text{ plane]}. \qquad (4.76)$$

The First Law of Thermodynamics states that $dU = T\, dS - p\, dV$. Since the loop integral of dU vanishes, we find that the area enclosed in the S--T plane above is exactly the work done in the loop process. In most cases in atmospheric science it is easier to deal with areas in the S--T plane than in the V--p plane. This is because the volume of a parcel is not easily observed, whereas its pressure (same as the environmental pressure and approximately equivalent to altitude) is easily observed. Similarly for adiabatic transformations $s = c_p \ln \theta$ is fixed and in a diabatic heating p is usually fixed.

Note that the rectangular shape of the closed figure for a Carnot cycle in the T--S plane does not depend on the system being composed solely of an ideal gas. Isotherms and adiabats are straight lines for any system in the S--T diagram in Figure 4.4. This means that the Carnot cycle is useful in describing cycles even of a composite system composed of any substances. A particular example is the case of a composite system consisting of a liquid in equilibrium with its vapor, both enclosed in a chamber. This last configuration is similar to a steam engine wherein vapor is condensed and later evaporated as separate legs of the cycle. It is shown in more advanced thermodynamics books that the Carnot cycle is the most efficient reversible cycle in terms of obtaining work from the system by extracting heat from a hot reservoir and rejecting part of it to a cold reservoir. Irreversible cycles are always less efficient than reversible ones. Thermodynamic loop diagrams for atmospheric processes will be exploited in later chapters as we analyze the energetics involved in parcels undergoing transitions in the real atmosphere.

Example 4.15: Carnot cycle for any system The S--T depiction of the Carnot cycle (Figure 4.4) works for any thermodynamic system. We can show a few properties that hold for the general case that were derived above only for the ideal gas case. For example, consider the work done in the cycle. The First Law tells us that since $(\Delta U)_{\text{loop}} = 0$ for the closed loop, $W_{\text{loop}} = Q_h - Q_c$. Next consider the Second Law which says that $(\Delta S)_{\text{loop}} = 0$; hence $Q_h/T_h = Q_c/T_c$. The efficiency is still W_{loop}/Q_h and it can be written as $1 - Q_c/Q_h = 1 - T_c/T_h$, exactly as for the ideal gas. Finally, we can verify that the area enclosed in the rectangle is W_{loop}. The area is given by

$$(\Delta T)(\Delta_{ab}S) = (T_h - T_c)(Q_h/T_h) = Q_h - T_c Q_h/T_c = Q_h - Q_c.$$

$\square$

4.6 Entropy summary

The extensive thermodynamic variables we have discussed so far are the volume V, the mass $\mathcal{M}$, internal energy U, the enthalpy H and the entropy S. There are also the intensive parameters, pressure p, and temperature T. It is best to think of the internal energy as a function of the entropy, volume and mass. Note that each is an extensive variable:

$$U = U(S, V, \mathcal{M}) \tag{4.77}$$

The following form of the First Law actually incorporates the Second Law (for fixed mass):

$$dU = T\,dS - p\,dV \tag{4.78}$$

and it shows explicitly that the internal energy is best characterized by these two variables, S and V. Note that this last expression shows how the internal energy changes in terms of other state variables (S and V). The associated intensive parameters are given by partial derivatives:

$$\left(\frac{\partial U}{\partial S}\right)_V = T, \quad \left(\frac{\partial U}{\partial V}\right)_S = -p \tag{4.79}$$

The enthalpy can be written:

$$H = H(S, p, \mathcal{M}) \tag{4.80}$$

and the First Law (combined with the Second) is expressed (for fixed mass) as

$$dH = T\,dS + V\,dp \tag{4.81}$$

along with the corresponding partial derivative expressions. We can also think of the entropy as the dependent variable:

$$S = S(U, V, \mathcal{M}) \tag{4.82}$$

and

$$dS = \frac{1}{T}\,dU + \frac{p}{T}\,dV \tag{4.83}$$

and

$$\left(\frac{\partial S}{\partial U}\right)_V = \frac{1}{T}, \quad \left(\frac{\partial S}{\partial V}\right)_U = \frac{p}{T} \tag{4.84}$$

4.7 Criteria for equilibrium

We return now to the extremum principle. If we consider a system and its surroundings, and we allow a spontaneous change to occur (release a constraint), we know that

$$dS_{\text{universe}} \geq 0 \implies dS_{\text{sys}} + dS_{\text{surr}} \geq 0 \qquad (4.85)$$

where the subscripts refer to the system and to the surroundings while the equality sign holds for a reversible process. Note that the entropy change for the surroundings can be written

$$dS_{\text{surr}} = \frac{dQ_{\text{surr}}}{T}. \qquad (4.86)$$

Here we have taken dQ_{surr} to be the same as $dQ_{\text{surr}}^{\text{rev}}$ since the surroundings might be assumed to undergo a reversible change (because of its large mass) while that in the system might not necessarily be reversible. But we can take $dQ_{\text{surr}}^{\text{rev}} = -dQ_{\text{sys}}$ since the heat gained by the surroundings has to be supplied by the system. The dQ_{sys} need not be reversible in this problem. We now can write

$$dS_{\text{sys}} - \frac{dQ_{\text{sys}}}{T} \geq 0 \qquad (4.87)$$

or rearranging to obtain the important formula

$$\boxed{dS_{\text{sys}} \geq \frac{dQ_{\text{sys}}}{T}} \quad \text{[equal for reversible, larger for irreversible].} \qquad (4.88)$$

The equality sign applies only if the process is actually reversible. As an example, consider the (irreversible!) free expansion studied in Example 4.11. In that case, $dQ_{\text{surr}} = 0$ while we found that $dS_{\text{sys}} > 0$.

Consider what happens to a system undergoing a spontaneous transition because some constraint has been relaxed or removed. We can use the First Law (which holds for reversible or irreversible transitions) to write

$$dS \geq \frac{dU + p\,dV}{T}. \qquad (4.89)$$

If the transition occurs at constant volume we can write

$$T\,dS \geq dU \quad \text{constant volume.} \qquad (4.90)$$

Similarly

$$T\,dS \geq dH \quad \text{constant pressure.} \qquad (4.91)$$

These can also be expressed as

$$\boxed{dS_{U,V} \geq 0, \quad dU_{S,V} \leq 0}$$ (4.92)

$$\boxed{dS_{H,p} \geq 0, \quad dH_{S,p} \leq 0}$$ (4.93)

where the subscripts indicate which variables are to be held constant. In words, if the internal energy and the volume are constrained to be fixed in the transition, the entropy of the system will increase in the transition. The other three inequalities can be expressed similarly.

4.8 Gibbs energy

There is another thermodynamic state function widely used in applications to atmospheric science, the *Gibbs energy* (sometimes called the *Gibbs free energy* or just the *free energy*). It proves to be useful for processes which occur at constant pressure and constant temperature. We can use the Gibbs energy to help us in deciding the direction of a chemical reaction and in determining the equilibrium phases or concentrations of chemical species in equilibrium. The Gibbs energy is particularly useful for open systems (those in which mass can enter or leave the system) and for systems in which the internal composition might change due to chemical reactions. We will take up some of these cases later in this chapter.

The Gibbs energy can be defined as

$$\boxed{G = H - TS} \quad \text{[definition of Gibbs energy]}$$ (4.94)

where H is enthalpy, T temperature and S is entropy. The differential of G can then be written:

$$dG = dH - T dS - S dT.$$ (4.95)

Substituting for dH:

$$dG = dQ + V dp - T dS - S dT.$$ (4.96)

Along a reversible path we can take $dQ = T dS$ which leads to

$$\boxed{dG = V dp - S dT} \quad \text{[differential for Gibbs energy].}$$ (4.97)

This last expression (which combines the First and Second Laws) tells us that G is a natural function of T and p, $G(T,p)$. Hence, in a change in which the mass, pressure and temperature are held fixed, the Gibbs energy will not change. This actually happens in a phase transition. For example, consider a chamber with a movable piston held at fixed temperature. Let the chamber contain a liquid with its own vapor in the volume above it. If the piston is withdrawn isothermally and quasi-statically some of the liquid will evaporate into the volume above the liquid surface. The pressure is just the vapor pressure and is constant since it depends only on the (fixed) temperature. We should note that the pressure in the liquid is the same as the pressure in the vapor (we ignore gravity here). Different positions of the piston (leading to different volumes of the vapor) correspond to the same temperature and the same pressure (in both liquid and vapor). Along this locus of points in the state space (say the V–p diagram) for this composite system the Gibbs energy is constant. We will return to the two-phase problem in Chapter 5.

Returning to the general problem we see from (4.97):

$$\boxed{\left(\frac{\partial G}{\partial p}\right)_{T,\mathcal{M}} = V, \quad \left(\frac{\partial G}{\partial T}\right)_{p,\mathcal{M}} = -S}$$
(4.98)

As indicated in an earlier chapter, the reactions of trace gases in the atmosphere occur at constant pressure and temperature. In this case the atmosphere which contains orders of magnitude more neutral background molecules (nitrogen, oxygen and argon) than the (usually trace) reactants acts as a massive thermal and pressure buffer holding the temperature and pressure constant. Hence, the reactions among trace gases in the atmosphere occur at fixed pressure (altitude) and temperature (that of the background gas). In these cases where T and p are held constant only the concentration of the species is allowed to change. This is the perfect setup for use of the Gibbs energy.

4.8.1 Gibbs energy for an ideal gas

Begin with the definition of G

$$G = H - TS.$$
(4.99)

Write $H = \mathcal{M}c_p T$ and the expression for entropy:

$$S = \mathcal{M}c_p \ln\left(\frac{T}{T_0}\left(\frac{p}{p_0}\right)^{-\kappa}\right)$$
(4.100)

where $\kappa = R/c_p$ as before. Then

$$G(T,p) = \mathcal{M}c_pT - \mathcal{M}c_pT \ln\left(\frac{T}{T_0}\left(\frac{p}{p_0}\right)^{-\kappa}\right)$$

$$= \mathcal{M}c_pT\left(1 - \ln\frac{T}{T_0}\right) + \mathcal{M}RT \ln\frac{p}{p_0}. \tag{4.101}$$

The *specific* Gibbs energy is G normalized by the mass, sometimes denoted $g(T,p)$:

$$\boxed{g(T,p) = c_pT\left(1 - \ln\frac{T}{T_0}\right) + RT \ln\frac{p}{p_0}}\ \text{[ideal gases]}. \tag{4.102}$$

Another form that is useful especially in chemistry is the *molar* Gibbs energy. Instead of specifying the specific Gibbs energy as per unit mass it may be more convenient to express it as a per mole quantity. In the expression for G note that for heating at constant pressure:

$$đQ = v\bar{c}_p\Delta T = \mathcal{M}c_p\Delta T \tag{4.103}$$

or

$$v\bar{c}_p = \mathcal{M}c_p \tag{4.104}$$

where c_p is in $J\,kg^{-1}\,K^{-1}$, $\mathcal{M}$ is in kg, v is the number of moles, and $\bar{c}_p$ is in $J\,mol^{-1}\,K^{-1}$. Also recall that R^* is the universal gas constant ($8.3145\,J\,mol^{-1}\,K^{-1}$).
Then

$$G(T,p) = v\bar{c}_pT\left(1 - \ln\frac{T}{T_0}\right) + vR^*T \ln\frac{p}{p_0} \tag{4.105}$$

and using $\overline{G}(T,p) = G(T,p)/v$ we have

$$\boxed{\overline{G}(T,p) = \bar{c}_pT\left(1 - \ln\frac{T}{T_0}\right) + R^*T \ln\frac{p}{p_0}}\ \text{[molar form]}. \tag{4.106}$$

In these expressions, κ is the same dimensionless number since $R/c_p = R^*/\bar{c}_p$.
A composite system consisting of several distinct subsystems of ideal gases $i = 1,\ldots,n$ leads to an expression for the Gibbs energy:

$$\boxed{G = \sum_{i=1}^{n} v_i \overline{G}_i(T_i,p_i) \left(= \sum_{i=1}^{n} \mathcal{M}_i g_i(T_i,p_i)\right)} \tag{4.107}$$

4.8.2 Equilibrium criteria for the Gibbs energy

As with entropy and internal energy, there is a condition for equilibrium for the Gibbs energy and it is often more useful than the others:

$$dS \geq \frac{dH - V dp}{T} = \frac{dG + T dS + S dT - V dp}{T} \tag{4.108}$$

which after dividing each side by dS simplifies to

$$0 \geq dG + S \, dT - V \, dp. \tag{4.109}$$

Now for a process at constant temperature and pressure,

$$\boxed{dG_{T,p} \leq 0} \text{ [equilibrium criterion for Gibbs energy]}. \tag{4.110}$$

This last is an important result, since so many processes take place at constant temperature and pressure. The inequalities derived earlier are somewhat less useful since in applications it is more difficult to control H, S or U. The last equation states that in a spontaneous process the (possibly composite) system will adjust its coordinates in such a way as to lower the value of the system's Gibbs energy; in this sense it behaves like a potential energy function in mechanics where a system tends toward minimum potential energy (see Figure 4.5). Equilibrium will establish itself at the minimum of $G(T, p)$, much as it did in the last chapter for a maximum of $S(U, V)$ only this time we have a function whose dependent variables are more under our control (or more to the point those found in naturally occurring circumstances).

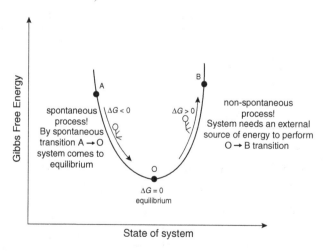

Figure 4.5 In a spontaneous transition the Gibbs energy is a minimum.

4.9 Multiple components

In all the thermodynamic functions we have studied so far we have ignored the fact
that the chemical composition of the system might change. In fact, the functions of
state might be summarized by:

$$U = U(S, V, \nu_1, \ldots, \nu_n) \tag{4.111}$$

$$H = H(S, p, \nu_1, \ldots, \nu_n) \tag{4.112}$$

$$G = G(T, p, \nu_1, \ldots, \nu_n) \tag{4.113}$$

$$S = S(U, V, \nu_1, \ldots, \nu_n) \tag{4.114}$$

where $\nu_1, \ldots, \nu_K$ indicate the number of moles of each chemical species in the
system. These molar indicators are thermodynamic coordinates. We can write

$$dG = \left(\frac{\partial G}{\partial p}\right)_{T,\nu_1,\ldots} dp + \left(\frac{\partial G}{\partial T}\right)_{p,\nu_1,\ldots} dT + \left(\frac{\partial G}{\partial \nu_1}\right)_{p,T,\nu_2,\ldots} d\nu_1 + \cdots . \tag{4.115}$$

We can write more compactly

$$dG = V dp - S dT + \overline{G}_1 d\nu_1 + \cdots + \overline{G}_n d\nu_n \tag{4.116}$$

where the $\overline{G}_i$ are the molar Gibbs energies for the individual components in the
mixture.[1] Note that a similar expression (but with differentials of their natural
variables serving as coefficients) holds for dU and dH with the same values of $\overline{G}_i$.

$$\left(\frac{\partial G}{\partial \nu_1}\right)_{p,T,\nu_2,\ldots} = \overline{G}_1, \quad \text{etc.} \tag{4.117}$$

expresses how much the composite Gibbs energy changes per mole of species 1
being added to the system. As with other intensive parameters in subsystems in
contact (such as T_1 and T_2 when the subsystems 1 and 2 are in diathermal contact,
or p_1 and p_2 when the pressures are allowed to be unconstrained) the specific Gibbs
energies $\overline{G}_i$ will tend toward equality when the number of moles of the different
species are allowed to vary (e.g., by chemical reactions or phase changes).

We can obtain some insight into this equalizing of the $\overline{G}_i$ by considering a system
at constant pressure and temperature in which there are two chemical species, A
and B. We have the reaction:

$$A \rightleftharpoons B. \tag{4.118}$$

---

[1] The $\overline{G}_i$ are denoted μ_i in the chemical literature and are called the *chemical potentials* of the system components.

The number of moles of B being created, $\delta\nu_B = -\delta\nu_A$. And since

$$dG = \left(\frac{\partial G}{\partial \nu_A}\right)_{T,p} d\nu_A + \left(\frac{\partial G}{\partial \nu_B}\right)_{T,p} d\nu_B = (\overline{G}_A - \overline{G}_B)\, d\nu_A. \tag{4.119}$$

We find that at equilibrium where $dG = 0$:

$$\overline{G}_A = \overline{G}_B. \tag{4.120}$$

The species A and B might be different phases of the same substance. We again find the equality of the two molar Gibbs energies for each phase when equilibrium is established. We will find this to be of great utility in the next chapter.

Next consider a system composed of two subsystems of equal volume, one is filled with ν_A moles of ideal gas species A the other with ν_B moles of B. Further, suppose the two gases have the same pressure p and temperature T. Now suppose the two subsystems are brought into material contact with the volume being the sum of the original volumes, the pressures and temperatures also being the same. What is the final Gibbs energy? What are the final enthalpy, internal energy, and entropy?

The initial Gibbs energy is:

$$G_{init} = (\nu_A + \nu_B)\overline{G}\,(p, T) \tag{4.121}$$

where we have used the same specific Gibbs energy $\overline{G}(p, T)$ for each of the ideal gases A and B. Once the gases are mixed into the larger volume, the total pressure will be the same, but the partial pressures will be only half as much since they occupy twice the volume but at the same temperature. Hence,

$$G_{final} = (\nu_A + \nu_B)\overline{G}\left(\frac{p}{2}, T\right). \tag{4.122}$$

Taking the difference and using the formula (4.101) we get

$$\Delta G = G_{final} - G_{init} = -(\nu_A + \nu_B)R^*T \ln 2 < 0. \tag{4.123}$$

This illustrates that the spontaneous process of mixing two ideal gases leads to a decrease in the Gibbs energy.

For the internal energy and enthalpy, the job is easy. The change of the internal energy is

$$\begin{aligned} \Delta U &= \nu_A \Delta\overline{U}_A + \nu_B \Delta\overline{U}_B \\ &= (\nu_A + \nu_B)\overline{c}_V \Delta T \\ &= 0, \end{aligned} \tag{4.124}$$

since $\Delta T = 0$. The same holds for enthalpy with the substitution $\bar{c}_V \rightarrow \bar{c}_p$.

As expected, during the mixing of ideal gases the Gibbs energy decreases, while the enthalpy and internal energies do not change.

To calculate the entropy change we choose a reversible isothermal path. We use

$$\Delta G = \Delta H - S\Delta T - T\Delta S \tag{4.125}$$

with $\Delta H = \Delta T = 0$. Hence,

$$\Delta S = -\Delta G/T = (\mathcal{M}_A R_A + \mathcal{M}_B R_B) \ln 2 > 0. \tag{4.126}$$

The mixing of two ideal gases causes an increase of the entropy as we learned earlier.

Suppose now that the gases A and B are identical, $\mathcal{M}_A = \mathcal{M}_B = \mathcal{M}$, $R_A = R_B = R$, $\nu_A = \nu_B = \nu$. Then, from (4.126) we get the increase of the entropy after mixing:

$$\Delta S = 2\mathcal{M}R \ln 2 \equiv 2\nu R^* \ln 2. \tag{4.127}$$

Does this make sense? In the beginning each subvolume contains the same number of moles of identical gases. What changes after the mixing of the gases? Nothing. Then the change in entropy should be zero. So we get two different answers for the same problem. This has become known as the Gibbs Paradox. The reason this paradox arises is that in classical physics we cannot consider the mixing of two identical gases as a limiting case of the mixing of two different gases. If we start our consideration for different gases, they have always to be different. It is impossible to get the answer for the entropy change of the mixing of two identical gases simply by equating the masses and the gas constants in equation (4.126). In classical physics the exchange of coordinates between two identical particles (gas molecules in our case) corresponds to a new microscopic state of the system (two gases in the cylinder), although nothing changes with such an exchange at the macroscopic level. This paradox does not exist in quantum theory, where the exchange of two identical particles does not correspond to a new microscopic state of the system. Therefore, when two identical gases are mixed, *the entropy does not change*.

Notes

Aside from the books already mentioned in earlier chapters, a beautiful treatment of thermodynamics from an axiomatic point of view is given by Callen (1985). Thermodynamics and its applications in engineering has a long history. A good introductory level engineering book is that by Çengal and Boles (2002). Both of

Emanuel's books (1994, 2005) as well as the book by Curry and Webster (1999) discuss the thermodynamics of convection phenomena.

Notation and abbreviations for Chapter 4

$\bar{c}_v, \bar{c}_p$	specific heats, the overbar indicates quantities expressed per mole ($J\,mol^{-1}K^{-1}$)
dQ_{rev}	infinitesimal absorption of heat, subscript indicating that the change be reversible (J)
$dS_{U,V}$	infinitesimal change in entropy during which U and V are held constant ($J\,K^{-1}$)
F	force (N)
g	Gibbs energy per kilogram ($J\,kg^{-1}$)
$\bar{G}$	Gibbs energy per mole or molar Gibbs energy ($J\,mol^{-1}$)
G	Gibbs energy (J)
H	enthalpy (J)
$\kappa = R/c_p$	(dimensionless)
$\mathcal{M}$	bulk mass (kg)
μ_i	chemical potential of species i, same as $\bar{G}_i$ ($J\,mol^{-1}$)
ν, ν_A, ν_B	number of moles, number of moles of species A, B
p	pressure (Pa)
$p(V)$	pressure as a function of volume; expression for a curve in the (p, V) plane
R	gas constant for a particular gas ($J\,kg^{-1}\,K^{-1}$)
s	entropy per unit mass (lower case indicates per unit mass) ($J\,K^{-1}\,kg^{-1}$)
S, S_A, S_B	entropy, entropy of state A, state B ($J\,K^{-1}$)
$S_{sys}, S_{surr}, S_{universe}$	entropy for the system, surroundings, universe (sys+surr)
θ	potential temperature (K)
U	internal energy (J)
U_A, U_B	internal energy at states A, B (J)
$W_{A\rightarrow B}, Q_{A\rightarrow B}$	work done by the system, heat taken into the system in going from state A to state B (J)

Problems

4.1 A parcel is lifted adiabatically from $z = 0$ to $z = H$, what is its change in entropy?

4.2 Compute the change in entropy for an ideal dry gas of mass $\mathcal{M}$ which is heated at constant volume from T_1 to T_2. Take $\mathcal{M} = 1\,kg$, $T_1 = 300\,K$ and $T_2 = 310\,K$.

4.3 A parcel is lifted isothermally from pressure p_0 to p_1. Find its change in potential temperature. Take $p_0 = 1000\,\text{hPa}$ and $p_1 = 500\,\text{hPa}$, $T_0 = 300\,\text{K}$.

4.4 A 1 kg parcel at $500\,\text{hPa}$ and $250\,\text{K}$ is heated with $500\,\text{J}$ of radiation heating. What is the change in its enthalpy? What is the change in its entropy? Its potential temperature?

4.5 A quantity 18 g of water is (a) heated from 273 K to 373 K, (b) evaporated to gas form, and (c) heated to 473 K. All steps are performed at constant pressure. Compute the change in entropy for steps (a), (b), and (c). Note: the heat capacity for water vapor is $\approx 2\,\text{kJ}\,\text{K}^{-1}\,\text{kg}^{-1}$.

Use these data in the next two problems: 2 kg of an ideal gas (dry air) is at temperature $300\,\text{K}$, $p = 1000\,\text{hPa}$.

Step 1: the volume is increased adiabatically until it is doubled.

Step 2: the pressure is held constant and the volume is decreased to its original value.

Step 3: the volume is held constant and the temperature is increased until the original state is recovered.

4.6 (a) Sketch the process (steps 1, 2 and 3) in the V–p plane.

(b) What are the volume and temperature at the end of step 1?

(c) What is the change of enthalpy ΔH, internal energy ΔU, and entropy ΔS, during step 1?

(d) How much work is done by the system in step 1?

4.7 Continuing Problem 4.6.

(a) How much work is performed in step 2?

(b) What is the total amount of work in all three steps?

(c) What is the entropy change ΔS in step 2?

(d) What are the total changes in U, H, S during all three steps?

4.8 A dry air parcel has a mass of 1 kg. It undergoes a process that is depicted in the V–p plane in Figure 4.6. Calculate the change of entropy, enthalpy and internal energy for this air parcel.

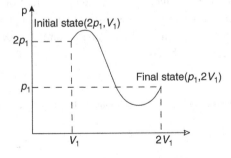

Figure 4.6 Diagram for Problem 4.8.

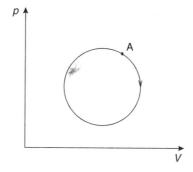

Figure 4.7 Diagram for Problem 4.9.

4.9 Find the change of internal energy and enthalpy for the cyclic process shown in Figure 4.7. Starting from point A describe how the temperature changes during this cyclic process.

4.10 Show that the work performed by a system during a reversible isothermal cycle is always zero.

4.11 Show that for the Carnot cycle of an ideal gas holds (4.74)

$$\frac{V_b}{V_a} = \frac{V_c}{V_d}.$$

Hint: Divide one of the four equations for work done along the different legs of the cycle by another one.

4.12 A tropical storm can be approximated as a Carnot cycle. Air is heated nearly isothermally as it flows along the sea surface (typically 27 °C). The air is lifted adiabatically in the eye wall to a height above the tropopause where it begins to cool due to loss of infrared radiation to space. The temperature where this occurs is about −73 °C. Finally the air descends to the surface adiabatically. Calculate the thermodynamic efficiency of this "heat engine."[2]

4.13 Show that the work done in a (reversible) Carnot cycle is the product of the entropy difference between the two adiabats and the temperature difference between the two isotherms (see Figure 4.4). The result holds for any system, not just an ideal gas.

4.14 Air is expanded isothermally at 300 K from a pressure of 1000 hPa to 800 hPa. What is the change in specific Gibbs energy?

---

[2] Kerry Emanuel (2005) explains this simple model at beginner's level in Chapter 10 of his book *Devine Wind*.

5

Air and water

In nature water presents itself in solid, liquid and gaseous phases. Energy transfers during transformations among these phases have important consequences in weather and climate. The system of redistribution of water on the planet constitutes the *hydrological cycle* which is central to weather and climate research and operations. Water is also an important solvent in the oceans, soils and in cloud droplets. The presence of tiny particles in the air can influence the formation of cloud drops and thereby change the Earth's radiation balance between absorbed and emitted and/or reflected radiation. These and other effects lead us into the fascinating role of water in the environment. Of course, thermodynamics is an indispensable tool in unraveling this very challenging puzzle.

5.1 Vapor pressure

We start with a discussion of the equilibrium gas pressure in a chamber in diathermal contact with a reservoir at a fixed temperature, T_0. The chamber is to have a volume that is adjustable, as shown in Figure 5.1. In the following let the chamber have no air present – only the gas from evaporation of the liquid. We are to choose a volume V such that there is some liquid present at the bottom of the chamber (we say here the bottom of the chamber, but otherwise we ignore gravity). There are gas molecules constantly striking the liquid from above and sticking. The rate at which particles enter the liquid phase will be proportional to the number density of molecules in the gas phase (recall from our discussion of kinetic theory that the flux of molecules per unit perpendicular area crossing a plane is $\frac{1}{4}n_0\bar{v}$ where $\bar{v}$ is mean speed of the vapor molecules (see Chapter 2)). Molecules in the liquid phase must have at least a certain minimum vertical component of velocity inside the condensed phase to escape the liquid surface (they have to overcome the potential energy necessary to leave the surface). If we wait for the equilibrium to establish itself, the rate of molecules leaving the liquid surface going into the volume above will exactly equal

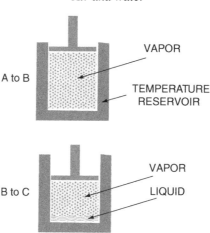

Figure 5.1 Quasi-static compression of a vapor at constant temperature. In A→ B, there is no liquid present. At point B, liquid begins to form on the base of the chamber. In B → C, there is liquid in equilibrium with the vapor.

the rate of molecules arriving and sticking. If the rate of departures should exceed the rate of sticking arrivals, the number density of gas molecules n_0 would steadily increase until the rates equalize.

If the volume of the chamber is decreased slightly, the equilibrium will have to be re-established. The steady state equality of arrival and departure rates can only be maintained for the same number density in the gas phase as before since the temperature is held fixed. In decreasing the volume we must condense a net amount of vapor molecules into the liquid phase under these conditions. As the excess number of sticking molecules falls into a potential energy well when they enter the liquid their velocities in the liquid increase (picture a slowly moving marble rolling off the table's edge, where its kinetic energy suddenly changes from near zero to a large value). This excess kinetic energy of the molecules entering the liquid is quickly shared with the other water molecules in the liquid, slightly raising its temperature. This tiny excess temperature over that of the reservoir in contact with the system is quickly wiped out (with a heat (enthalpy) transfer to the reservoir), maintaining the isothermal constraint. As the volume is decreased, there is another form of energy being added to the system, because during the compression, work is being performed on the system by the piston (maintaining constant pressure).

Consider the situation in a V–p diagram (Figure 5.2). We wish to trace an isotherm for this system. We start at point A where all the matter in the chamber is in the gas phase. We compress the gas isothermally until we reach point B where liquid begins to condense on the floor of the chamber (no droplets, please, because surface tension

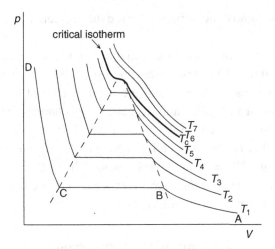

Figure 5.2 Pressure versus volume diagram for a mixture of liquid and vapor. We start at point A where the system is all vapor, and compress isothermally to point B, where liquid begins to appear. In going from B to C a mixture of liquid and vapor is present. Along this second stage the isotherm is also an isobar. The portion of the curve to the left of C represents the liquid phase. The critical isotherm is shown by the bold line. $T_1 < T_2 < T_3 < \ldots < T_7$. The dashed line bounds the area where the liquid and its vapor are in equilibrium.

of the curved droplet surfaces would complicate the energetics here; but also no gravity in this experiment – the water could congregate on the ceiling for all we care here). As we isothermally and quasi-statically compress further, the pressure remains constant as more matter is converted from gas to liquid phase. Heat released from the condensation and from the work performed during the compression must be transferred from the chamber to the reservoir in order to maintain the same temperature. The change in internal energy is composed of two contributions, the work done by the piston on the gas and the heat associated with the matter being converted from vapor to liquid. The change in enthalpy of the system does not depend on volume, so its change only involves the condensation contribution. The change in enthalpy in moving along from B to C in Figure 5.2 is

$$\Delta H = Q = -(\Delta \mathcal{M}_\ell) L \tag{5.1}$$

where $\Delta \mathcal{M}_\ell$ is the (positive) amount of matter condensed in the process and $L = \Delta H_{\text{vap}}$ is called the *enthalpy of vaporization* (*latent heat of evaporation* in old fashioned terminology). For water at $0\,^\circ\text{C}$, $L = 2.500 \times 10^6\,\text{J kg}^{-1}$, and it is nearly independent of T (error $< 1\%$, over the range of interest in atmospheric science).

When the volume is reduced to such an extent that we have only liquid in the chamber, the further decrease in volume requires a very large increase in pressure, because liquids are almost incompressible (see the segment of the curve C → D of the isotherm).

If we plot isotherms corresponding to higher temperatures ($T_1 < T_2 < T_3 \ldots$), we see that the length of the horizontal portion of the isotherm decreases. This means that with increasing temperature the volume interval for which the liquid and vapor can coexist in equilibrium decreases. This happens until we reach the so-called *critical isotherm*, where this interval shrinks to a point. The temperature corresponding to this isotherm is called the *critical temperature*. At the critical temperature T_c, $\partial^2 p/\partial V^2 = 0$, an *inflection point*. The isotherms with temperature well above the critical temperature are hyperbolae, because the substance at very high temperatures behaves like an ideal gas.

5.2 Saturation vapor pressure

The equilibrium pressure of water vapor above a flat surface of liquid water in a chamber such as shown in Figure 5.1 is called the *saturation vapor pressure*. It is independent of the shape of the volume in the cylinder (since it only depends on the number density n_s of the vapor). The saturation vapor pressure (usually denoted e_s) is however a very strong function of temperature T. This is intuitively reasonable since an increase in temperature will increase the proportion of liquid molecules having velocities above the threshold to depart from the surface. More departures will require more arrival rates to maintain equilibrium. This in turn will require a larger number density which is proportional to the vapor pressure. Note that the flux of molecules moving down perpendicularly is $\frac{1}{4}n_s\bar{v}$ (see Chapter 2).

Figure 5.3 shows a graph of the saturation vapor pressure of water over a flat liquid surface versus the temperature in degrees Celsius. Many aspects of weather and climate depend on this very rapid increase with temperature. As a rough but

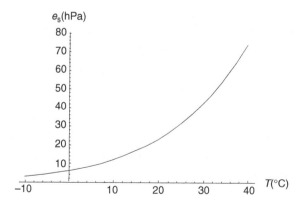

Figure 5.3 Saturation vapor pressure for water over a flat liquid surface versus temperature in degrees Celsius.

useful rule of thumb, the saturation vapor pressure doubles for every $10\,°C$ increase in temperature (at least in the range of interest for atmospheric science). Even so, at moderate temperatures the saturation vapor pressure is very small compared to atmospheric pressure near the surface (usually 5 to 30 hPa compared to 1000 hPa).

Does the presence of dry air affect the saturation vapor pressure of water? Perhaps the added pressure of the air on the liquid surface squeezes more water molecules into the vapor phase. But on the contrary, some air dissolves in the liquid and thereby might hinder the flux of molecules out of the liquid surface. Both effects are present but together their impact is less than 1% of the saturation vapor pressure.

5.3 Van der Waals equation

As we learned earlier, the approximation of an ideal gas works well if we can neglect the intermolecular forces. This is virtually always the case for the major constituents of air at Earth-like conditions. But as a gas nears its critical temperature and the liquid or solid state can coexist with the gas phase, the departure from ideality is important. As we see from Figure 5.2 the ideal gas equation of state describes the behavior of real gases in limiting cases of high temperatures and low pressures. Isotherms for an ideal gas are rectangular hyperbolae ($p \propto 1/V$). A small pressure decrease leads to a large increase in volume (B to A in Figure 5.2). However, the ideal gas equation of state is no longer a good approximation when the temperature of the gas is below its critical point, and the volume is in the range where the isotherms become horizontal (see the flat segment C to B in Figure 5.2); i.e., there is a mixture of liquid and gas in equilibrium together.

A very useful equation that describes the behavior of many substances over a wide range of temperatures and pressures was derived by van der Waals. The van der Waals equation for 1 mol of gas is:

$$\boxed{\left(p + \frac{a}{v^2}\right)(v - b) = R^*T}\quad \text{[van der Waals equation]}\qquad(5.2)$$

where a and b are constants (different for different substances) and v is the volume per mole of the gas (i.e., the *reduced volume* or the *specific volume*). The term b in (5.2) is due to the finite size of the molecules, while the term a/v^2 is due to the effect of the attractive molecular forces. For $a = b = 0$, the van der Vaals equation reduces to the Ideal Gas Law (5.2).

Usually the van der Waals equation is written in the form

$$p = \frac{R^*T}{v - b} - \frac{a}{v^2}.\qquad(5.3)$$

Figure 5.4 shows an example of isotherms calculated using the van der Waals equation. If we compare Figures 5.2 and 5.4, we see that van der Waals isotherms

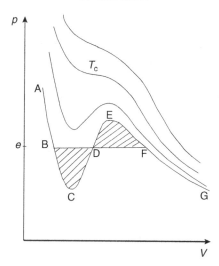

Figure 5.4 Van der Waals isotherms. The isotherm with the inflection point is the critical isotherm. The equilibrium vapor pressure e is such that the shaded areas are equal.

reproduce many features of real gas behavior. As shown in Figure 5.4 for large v and low p there is a large increase in volume with a small decrease in pressure. For a liquid (small v and high p) there is a small decrease in volume with a large increase in pressure. There is a critical isotherm with temperature $T = T_c$ indicating a point of inflection ($\partial^2 p/\partial v^2 = 0$). The isotherms with temperatures higher than T_c are very similar to those in Figure 5.2. However, the isotherms with temperature less than T_c look very different: they are not horizontal in the region where two phases, water and vapor, coexist. Consider one particular isotherm ABCDEFG derived from the van der Waals equation. Let us compress the gas until saturation occurs at point F on the isotherm. Then with a further decrease of the volume there is no increase in pressure, which corresponds to the horizontal stretch FB. Instead, the van der Waals isotherm shows an increase in pressure (part of the diagram FE). Along this branch of the curve the vapor is *supersaturated*. Vapor can theoretically exist for these values, but if a small impurity is present such as a dust particle, or a scratch on the wall, the vapor will begin to condense on this site and the system will collapse to the flat horizontal line BF in Figure 5.4. In other words this state of the vapor is unstable: any disturbance causes it to migrate to a stable condition which contains two subsystems, vapor and liquid. So, if we plot the van der Waals isotherm for a given temperature, we will not find the flat portion (BDF) which we know should be there (from experiment). We have to put it in "by hand." But how do we decide the proper pressure value at which to insert this flat portion? The rule (first discovered by Maxwell) is that the areas bounded by the curves BCDB and DEFD have to be equal. Let us sketch a proof. Consider the cycle FEDCBDF in Figure 5.4

(a "figure 8" on its side). From the First Law we know that for an isothermal process the work done during a closed cycle is equal to the amount of heat absorbed by the system, $\Delta W = \Delta Q$, since the change in internal energy is zero for a cyclic process. We also know that the loop integral $\oint dQ_{rev}/T = 0$ (our process is contrived to be reversible). For an isothermal process we can take temperature out of the integral and get $Q_{loop} = 0$. Since $Q_{loop} = 0$, we also have $W_{loop} = 0$. If the horizontal line were not such that the areas are equal, our imaginary (but realizable) process would violate the laws of thermodynamics (either $\oint dU \neq 0$ or $\oint dS \neq 0$ or both).

An excellent discussion of unstable states (supersaturated, etc.) can be found in advanced books, especially the discussion in Callen (1985), where the case is illustrated with the van der Waals system.[1] The criterion for stability can be expressed in terms of the concavity or convexity of the thermodynamic functions:

$$\frac{\partial^2 S}{\partial U^2} \geq 0, \quad \frac{\partial^2 S}{\partial V^2} \geq 0 \quad \text{[stability criterion]} \tag{5.4}$$

or for the Gibbs energy:

$$\frac{\partial^2 G}{\partial T^2} \leq 0, \quad \frac{\partial^2 G}{\partial p^2} \leq 0 \quad \text{[stability criterion]}. \tag{5.5}$$

If the graphs for $S(U, V, \mathcal{M})$ and $G(T, p, \mathcal{M})$ have the wrong sign of concavity the branch of the curve where the criteria fail will be unstable.

5.4 Multiple phase systems

We proceed with the case of water in both its liquid and vapor forms in equilibrium in a container. This is a one-component (only one chemical species is present) system with two phases (liquid and gas) in equilibrium. Experience tells us that the two phases can coexist in equilibrium in this configuration. In fact, we have seen that for a given mass of the substance there is a range of values of volume for which the equilibrium exists with transfers of mass from one phase to the other as the volume is changed (at constant pressure and temperature). This is the horizontal line CB in Figure 5.2. Let the temperature and pressures be T_0 and p_0 along CB. In the T–p plane this line is a single point (T_0, p_0), see Figure 5.5. If we were to make an infinitesimal change in the temperature reservoir to $T_0 + \Delta T$ (see Figure 5.6), then we would move to a higher horizontal line in Figure 5.2, thereby operating at

---

[1] Callen gives an expression for the entropy of a van der Waals gas: $S(u, v) = vR^* \ln \left((v - b)(u + a/v)^c\right) + vs_0$ where c is the molar heat capacity at constant pressure. For water vapor, $a = 0.544$ Pa m^6, $b = 30.5 \times 10^{-6}$ m^3, and $c = 3.1$.

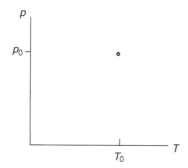

Figure 5.5 A point in the T–p plane in which liquid and vapor are in equilibrium.

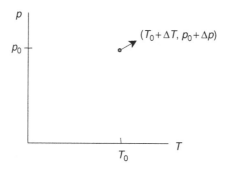

Figure 5.6 As the temperature is increased from T_0 to $T_0 + \Delta T$, the saturation vapor pressure will increase from p_0 to $p_0 + \Delta p$.

$(p_0 + \Delta p, T_0 + \Delta T)$. As we change from one flat line in the V–p plane, we trace out a new curve in the T–p plane. Let us call it $p_{\text{equil}}(T)$.

Along this curve, $p = p_{\text{equil}}(T)$, the two phases can exist in equilibrium. In fact, if T and p lie on the curve (i.e., $p = p_{\text{equil}}(T)$) then the volume can be varied isothermally and isobarically causing mass to transfer from one phase to the other until one of the phases is exhausted. The point in Figure 5.5 lies between points B and b in Figure 5.7. The variation can be thought of as into or out of the T–p plane along the V (volume) axis.

The upshot of all this is that when the two phases are together in equilibrium there will be a unique curve in the T–p plane. This line is of great interest to us. For example its slope tells us how much the saturation vapor pressure will increase for a small change in the temperature.

Water can form ice as its solid phase. It turns out that a single-component system such as pure water can coexist in all three phases simultaneously only at a single point in the phase diagram called the *triple point*. The triple point for water is 273.16 K at a pressure of 6.11 hPa. At pressures below 6.11 hPa ice and vapor can

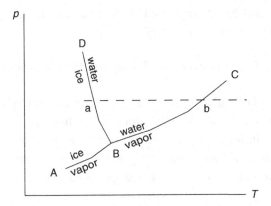

Figure 5.7 A schematic phase diagram in the $T–p$ plane of the phases of water. Below the line ABC the phase is vapor. To the left of ABD the phase is solid. Above DBC the phase is liquid. These lines are called phase boundaries since along them two phases can coexist. The point B is the so-called triple point since all three phases can coexist at this point. The dashed line ab indicates atmospheric pressure. The boiling point is at b.

coexist, but no liquid water will exist in equilibrium. There are many other phases of water in its different crystalline forms, and these are important in high pressure situations well outside the range encountered in atmospheric applications.

5.5 Phase boundaries

Consider a point representing a state in the $T–p$ plane. At such an arbitrary point the Gibbs energy for each of the subsystem phases will have given values $\mathcal{M}_l g_l(T, p)$ and $\mathcal{M}_g g_g(T, p)$, where $\mathcal{M}_l$ and $\mathcal{M}_g$ are the masses of the liquid and gaseous subsystems and $g_l(T, p)$ and $g_g(T, p)$ are the specific Gibbs energies for each (by use of the term *specific* we mean that each is per unit mass). The Gibbs energy for the whole system is the sum of the mass weighted specific Gibbs energies. Note that when the two phases coexist in equilibrium, the pressure, temperature and specific Gibbs energies of each phase are homogeneous throughout (e.g., the specific Gibbs energy of the liquid equals that of the vapor). Only the density differs from one phase to the other. To show this, suppose we are on a phase boundary in the $T–p$ plane (the curve of equilibrium states described in the last section), and we change the volume of the system slightly, ΔV. In that transition, the total system Gibbs energy, $G(T, p)$ does not change since p, T, and $\mathcal{M}$ do not change. This means that

$$\Delta \mathcal{M}_l g_l(T, p) + \Delta \mathcal{M}_g g_g(T, p) = 0 \tag{5.6}$$

where we have made use of the fact that the specific Gibbs energies do not change because p and T are fixed. Now in the last equation we can note that $\Delta \mathcal{M}_l = -\Delta \mathcal{M}_g$

because the mass has to be conserved. This leads to the interesting and useful conclusion that along a phase boundary,

$$\boxed{g_l(T,p) = g_g(T,p)} \quad \text{[along a phase boundary]}. \tag{5.7}$$

In the T–p phase plane different regions represent different phases of this two-phase system. A phase boundary exists along which the liquid and gaseous phases can coexist in equilibrium. We have shown above that the specific Gibbs energies for each individual phase are equal along the phase boundary in the T–p plane. This result will allow us to calculate the slope of the phase boundary in the next section.

Gibbs phase rule In the last two sections we discussed multiple phases and in particular the case of water and its three phases. In general there might be more than one component as well (for example, a mixture with different phases for each). The intensive variables in the problem are the temperature and the pressure (common and equal for all the components). We also know that when the system is in equilibrium, the specific Gibbs energies for a given component $\overline{G}_i(T,p)$ will be equal for the phases of that component. In looking back at the water problem we see that there are regions in the T–p diagram where both T and p can be varied independently. These are regions where there is a single phase present. The lines in the diagram (Figure 5.7) represent a locus of points where two phases are present in equilibrium. Finally, the triple point is the single point where three phases are present in equilibrium. This is the situation when there is only one component present (water).

The *number of degrees of freedom* denoted here as F (different from the same name used in kinetic theory) refers to the number of ways one of the intensive variables $(T, p, \overline{G}_1, \overline{G}_2, \dots, \overline{G}_c$, where C is the number of components) can be varied independently. For example, in the regions away from phase boundaries in Figure 5.7 both T and p may be varied independently (two degrees of freedom), but on a phase boundary, only one of these variables is independent since the phase boundary is defined by a function, $p = p(T)$ (one degree of freedom). At the triple point the number of degrees of freedom is zero.

It is possible to derive a formula for the number of degrees of freedom for a multi-component, multi-phase system and it is worth presenting here. First note that the number of molar concentration variables is C, but only $C - 1$ are independent, since we are interested only in mole fractions. The number of phases is P. So the total number of these intensive variables is $P(C - 1)$. But there are some relations between these variables because some of the Gibbs energies are related to one another. For each individual component the Gibbs energies of the phases have to be equal. For each component there are $P - 1$ relations. (For example, if there are two phases there is only one relation, say $\overline{G}_1 = \overline{G}_2$, etc.) This reduces the number of independent intensive variables by $C(P-1)$. We still have the pressure and temperature that are independent to add. Thus we obtain

$$F = P(C-1) - C(P-1) + 2 \qquad (5.8)$$

leading to the *Gibbs phase rule*:

$$\boxed{F = C - P + 2} \quad \text{[number of degrees of freedom: Gibbs phase rule].} \qquad (5.9)$$

We see that for a single-component system with one phase, $F = 2$ (the regions between phase boundaries in Figure 5.7). When there are two phases present (on a phase boundary line), $F = 1$. When all three phases are present, $F = 0$, the triple point.

Other systems of interest include the case of water with a dissolved solute such as salt. This would be a two-component system with two phases (the liquid solution and the saturated vapor in equilibrium with it above). This case will be discussed later.

5.6 Clausius–Clapeyron equation

Having established that the specific Gibbs energies for liquid and gaseous phases are the same along the phase boundary, we can now proceed to calculate the slope of the phase boundary in the T–p plane. This slope is the rate of change of the vapor pressure with respect to temperature as the system is allowed to move along the phase boundary. This slope measures the rate at which the saturation vapor pressure increases for incremental changes in temperature – an important quantity in meteorology.

First consider such a reversible change of the composite system (gas and liquid in equilibrium) along the phase boundary. We have:

$$\Delta g_l(T,p) = \Delta g_g(T,p). \qquad (5.10)$$

Then (using infinitesimal notation instead of Δ)

$$-s_l \, dT + v_l \, dp = -s_g \, dT + v_g \, dp \qquad (5.11)$$

where the small letters s and v refer to *specific* entropy and volumes respectively, and henceforth we denote the saturation vapor pressure as e_s. Rearranging:

$$\frac{de_s}{dT} = \frac{s_g - s_l}{v_g - v_l}. \qquad (5.12)$$

First, notice that $v_g \gg v_l$ (e.g., for one gram mole of vapor 22.4×10^3 cm^3 versus 18 cm^3 of water) so that v_l can be neglected. The specific volume of an ideal gas is RT/p, where R is the gas constant for the particular species (here water vapor). The difference of specific entropies can be calculated. This difference is the change in specific entropy as we convert a unit mass of liquid into gas form at a fixed temperature: $\Delta H_{vap}/T = L/T$, where L is the enthalpy of vaporization (per kilogram). We arrive at the *Clausius–Clapeyron equation*:

$$\boxed{\frac{de_s}{dT} = \frac{Le_s}{RT^2}} \quad \text{[Clausius–Clapeyron equation]}. \qquad (5.13)$$

Before proceeding to integrate this equation to find an expression for $e_s(T)$, it should be noted that the procedure just employed is very general and can be applied to many other problems. While we will not pursue it here, it is perhaps clear that the equilibrium we speak of could be that of chemical species instead of phases, or it could be a combination of both. In physical chemistry texts the technique of equilibrium boundaries utilizing the Gibbs energy can be found to lead to such diverse rules as the temperature dependence of reaction rate coefficients.

5.7 Integration of the Clausius–Clapeyron equation

To proceed we divide each side of the equation by e_s and multiply through by dT. The left-hand side will be a function only of e_s and the right-hand side will be only a function of T. This allows us to integrate:

$$d \ln e_s = \frac{L\,dT}{R\,T^2} \qquad (5.14)$$

$$\Rightarrow \quad \ln \frac{e_s}{e_s(0)} = \frac{L}{R}\left(\frac{1}{T_0} - \frac{1}{T}\right). \qquad (5.15)$$

Next we choose the lower limit to be 273.2 K so that $e_s(0) = 6.11$ hPa. The value 6.11 hPa has to come from observations – thermodynamics cannot tell us the value of such a constant. After all, this *integration constant* should be different for different substances (e.g., compare this value to the vapor pressure of mercury at $20\,°C$ which is 0.16 Pa). Inserting the other numerical values leads to:

$$\ln \frac{e_s}{6.11\text{ hPa}} = \frac{L_{vap}}{R_w}\left(\frac{1}{273.2} - \frac{1}{T}\right) = 19.83 - \frac{5417}{T} \qquad (5.16)$$

where we have inserted the numerical values for $R = R_w$ and $L = L_{vap}$. This equation can also be rearranged to give the handy formula:

$$\boxed{e_s = 2.497 \times 10^9\, e^{-5417/T} \ \ (\text{hPa})} \quad \text{[integrated form of the}$$

Clausius–Clapeyron equation]. $\qquad (5.17)$

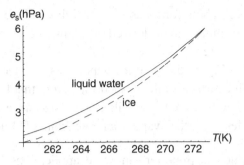

Figure 5.8 Saturation vapor pressure over ice (dashed line) and liquid water (solid line). The curves were computed with the Clausius–Clapeyron equation with $L_{vap} = 2.5 \times 10^6$ J kg^{-1} and $L_{sublime} = 2.83 \times 10^6$ J kg^{-1}.

The resulting graph is shown in Figure 5.8. If we consider the vapor to be in equilibrium with an ice surface we must use $L_{sublime} = 2.83 \times 10^6$ J kg^{-1} and the result of this is shown in Figure 5.8 as the dashed line. Note that below $T = 273$ K the saturation vapor pressure is larger over liquid than over ice. This means that if there is an ice surface in the chamber, the ice surface will not be in equilibrium with the vapor in the chamber, which is at the saturation value for a liquid surface. The upshot is that the ice volume will grow in size at the expense of the liquid mass. Eventually the vapor pressure in such a chamber will become that of the saturation vapor pressure over ice. This effect is important in a cloud at temperatures below freezing (0 °C) in which ice crystals are embedded in a field of *supercooled* water droplets. The term supercooled is applied since water droplets can be below the freezing point (0 °C) without actually freezing. We will see later that the presence of an impurity in the droplet, such as silver chloride, can cause the droplet to freeze at slightly higher temperatures. The point at which all the supercooled droplets freeze is −40 °C.

Example 5.1: comparing vapor pressure over ice and liquid water A plot of the saturation vapor pressure over a flat liquid water surface is shown in Figure 5.8. If we consider the vapor to be in equilibrium with an ice surface we must use $L_{sublime} = 2.834 \times 10^6$ J kg^{-1}. Then the vapor pressure over ice is approximately

$$\ln\left(\frac{e_s^{ice}}{6.11\ \text{hPa}}\right) = 22.50 - \frac{6148}{T}. \tag{5.18}$$

The corresponding formula for vapor pressure over a liquid surface is

$$\ln\left(\frac{e_s^{water}}{6.11\ \text{hPa}}\right) = 19.83 - \frac{5417}{T}. \tag{5.19}$$

where the temperature in both formulas [2] is in kelvins and the vapor pressure is in hPa. More accurate formulas can be derived by taking into account the temperature dependence of L_{sublime}, etc.

These are shown in Figure 5.8 with the vapor pressure over liquid as the solid line and over the solid surface as the dashed line. Note that below $T = 273$ K, if there is a piece of ice in the chamber, it will grow in size since the evaporation rate from the ice will be less than the evaporation rate from a flat liquid surface. ☐

Another effect which is important in some applications is the difference in saturation vapor pressures for different isotopes of water, H_2O^{18} versus H_2O^{16}. Both of these isotopes are radiologically stable (they do not decay) and both are found in nature. The heavier isotope makes up only about 0.20% of oxygen atoms. The vibration frequencies of the water molecules are affected slightly by the small amount of the heavier isotope. This leads to a very small change in the saturation vapor pressure of water having more or less of the heavy isotope present in the liquid. This small effect has a temperature dependence and it leads to slightly different evaporation rates over warm versus cool ocean waters. This leads to a different concentration of the isotope ratio in the water vapor over these different types of ocean water. The ratio of the heavy isotopic water to the lighter one can be measured in ice cores and in other material. The application is in paleoclimatology where snow deposited on polar ice fields leaves in its layering a record of temperature signatures of past climates. This is a very active area of current research – not just for the water isotopes but for those of many other elements.

5.8 Mixing air and water

When there is a mixture of air and water vapor the effective molecular weight of the gas changes slightly. We can use Dalton's Law (Section 2.5) to find the effective value of the gas constant, R_{eff}. First we find the effective value of the molecular weight when some water vapor is present. The result is [3]

---

[2] Emanuel (1994) contains extended discussions of these relations.
[3] The algebra required begins here and continues on the next page.

$$\frac{1}{M_{\text{eff}}} = \frac{1}{\mathcal{M}_v + \mathcal{M}_d}\left(\frac{\mathcal{M}_v}{M_w} + \frac{\mathcal{M}_d}{M_d}\right)$$

$$= \frac{\mathcal{M}_d/M_d}{\mathcal{M}_v + \mathcal{M}_d}\left(1 + \frac{\mathcal{M}_v/M_d}{M_w/M_d}\right)$$

$$= \frac{1}{M_d}\left(\frac{1}{1 + \mathcal{M}_v/\mathcal{M}_d}\right)\left(1 + \frac{\mathcal{M}_v/M_d}{M_w/M_d}\right)$$

$$\frac{1}{M_{\text{eff}}} \approx \frac{1}{M_{\text{d}}}(1 + 0.60w) \qquad (5.21)$$

where w, the *mixing ratio*, is given by the ratio of the mass of water vapor to the mass of dry air in a parcel:

$$w = \frac{M_{\text{v}}}{M_{\text{d}}}. \qquad (5.22)$$

Entering the calculation was the ratio of molecular weights of water to air:

$$\frac{M_{\text{w}}}{M_{\text{d}}} = \frac{18.02}{28.97} = 0.622. \qquad (5.23)$$

The *mixing ratio* w is usually given in units of grams of water vapor per kilogram of dry air (of course, in equations such as those just above it would be in (kg vapor) (kg air)$^{-1}$).

The effective value of the gas constant is then

$$R_{\text{eff}} = \frac{M_{\text{d}}}{M_{\text{eff}}}R_{\text{d}} = \frac{28.97}{M_{\text{eff}}}287 = \frac{8314}{M_{\text{eff}}}. \qquad (5.24)$$

The results above allow us to write the equation of state for the moist air as

$$p = \rho R_{\text{eff}} T = \rho R_{\text{d}}(1 + 0.60w)T = \rho R_{\text{d}} T_{\text{v}} \qquad (5.25)$$

where $R_{\text{d}} = 287 \, \text{J} \, \text{kg}^{-1} \, \text{K}^{-1}$ and

$$\boxed{T_{\text{v}} \equiv (1 + 0.6w)T} \quad \text{[virtual temperature]} \qquad (5.26)$$

is called the *virtual temperature*. Note that the virtual temperature is always larger than the actual temperature, but that they seldom differ by more than 1 K (w is seldom greater than 4×10^{-2} (kg vapor) (kg dry air)$^{-1}$.

The use of virtual temperature allows the meteorologist (whose interest is buoyancy) to correct the density to lower values when water vapor is present while retaining the simplicity of the Ideal Gas Law for dry air ($R_{\text{d}} = 287 \, \text{J} \, \text{kg}^{-1} \, \text{K}^{-1}$). This

Footnote 3 *continued*

$$= \frac{1}{M_{\text{d}}}\left(\frac{1}{1+w}\right)\left(1 + \frac{w}{0.622}\right)$$

$$\approx \frac{1}{M_{\text{d}}}(1 - w + w^2 + \cdots)(1 + 1.607w)$$

$$\approx \frac{1}{M_{\text{d}}}(1 + 0.60w). \qquad (5.20)$$

works to a very good approximation in practical situations. Remember that water vapor has a lower molecular weight but each water vapor molecule at temperature T has the same effect on pressure as an air molecule. Thus if there is a mixture of water vapor and air where the total pressure is the same, the density will be lower than for the same volume of dry air at the same temperature and pressure.

The *saturation vapor mixing ratio* is denoted w_s and it is a strong function of temperature. Keep in mind that the saturation vapor mixing ratio is also a function of the air pressure (equivalent to altitude) because it is the ratio of water mass to air mass.

The *relative humidity* is given by

$$\boxed{r = \frac{w}{w_s}} \quad \text{[relative humidity]}. \tag{5.27}$$

Or in terms of percent:

$$RH(\%) = \frac{w}{w_s} \times 100. \tag{5.28}$$

The *dew point* temperature, T_D, is the temperature at which

$$\boxed{w = w_s(T_D)} \quad \text{[dew point]}. \tag{5.29}$$

In other words, for a given value of w, it is the temperature for which that value of mixing ratio, w, is equal to the saturation mixing ratio, w_s. We reach the dew point by cooling at constant pressure to the temperature where w in a parcel reaches its saturation value.

To find the relationship between the partial pressure due to vapor and the partial pressure of the dry air in a parcel, we write:

$$e = \mathcal{M}_v R_w T / V \tag{5.30}$$

$$p = \mathcal{M}_d R_d T / V. \tag{5.31}$$

Taking the ratio and rearranging:

$$\frac{e}{p} = \frac{\mathcal{M}_v}{\mathcal{M}_d} \frac{M_d}{M_w} \tag{5.32}$$

or

$$\boxed{w = \epsilon \frac{e}{p}} \quad \text{[mixing ratio in terms of vapor and air pressure]} \tag{5.33}$$

where

$$\boxed{\epsilon = \frac{M_w}{M_d} = 0.622} \tag{5.34}$$

And of course the formula holds for the saturation case as well

$$w_s = \epsilon \frac{e_s}{p} \quad \text{[saturation mixing ratio]}. \tag{5.35}$$

Strictly speaking, p above is p_{dry} which is not $p_{atmo} = e + p_{dry}$, but usually the difference is negligible.

We can also write

$$w = w_s(T_D) = \epsilon \frac{6.11\,(hPa)}{p} \exp\left(\frac{L}{R_w}\left(\frac{1}{273.2} - \frac{1}{T_D}\right)\right) \tag{5.36}$$

and since we are to lower the temperature to T_D isobarically we can say

$$e = e_s(T_D) = 6.11\,(hPa)\exp\left(\frac{L}{R_w}\left(\frac{1}{273.2} - \frac{1}{T_D}\right)\right). \tag{5.37}$$

Then if either w or e are known we can find T_D by solving one of these equations for T_D.

Example 5.2 Water vapor is mostly distributed in the boundary layer of the atmosphere (lowest 1–2 km). Assume the atmosphere is isothermal at 300 K and that the surface humidity is 95%. If all the vapor is distributed uniformly in the first 1.5 km, how much water lies above a given square meter of surface in $kg\,m^{-2}$? Also express the result in mm equivalent of liquid water.

Answer: First compute the saturated water vapor pressure from the Clausius–Clapeyron equation: $e_s = 36\,hPa$. Then the saturation mixing ratio is given by $w_s = (0.622) \times (0.036)$ (kg water) (kg air)$^{-1} = 0.022$(kg water) (kg air)$^{-1}$. Thus, the water vapor mixing ratio is 95% of this, 0.021 (kg vapor) (kg air)$^{-1}$. The mass of air in the 1.5 km column is $1500\,m^3 \times 1.2\,kg\,m^{-3} = 1800\,kg$. Multiplying yields 38 kg of water vapor in the 1 m^2 column. The equivalent depth of liquid water is $\mathcal{M}_{water}/(\rho_{liq\;water} \times 1\,m^2) = 38$mm. □

Example 5.3: dry line This is a fairly sharp boundary often found in the area east of the Rocky Mountains running north–south. The boundary separates dry air on the west from moist air on the east. The dry air can come from winds from the south (Mexico) or dry air descending from the Rockies. To the east the air may be moist because of southerly flow from the Gulf of Mexico. Reductions of dew point of as much as 18 °C can be found in going from east to west across the dry line. If the air column is pushed eastwards the heavier dry air can wedge under the lighter moist air and sometimes lead to cloudiness or even rain. □

Another measure of moisture encountered in atmospheric science is the *specific humidity, q*. It is the number of grams of water vapor per unit mass of air plus water vapor (usually expressed in kilograms). In applications, q is numerically close enough to the mixing ratio w that one seldom needs to distinguish between them.

Example 5.4 Compare q and w if the pressure is 800 hPa and $q = 0.010$ (kg vapor) (kg moist air)$^{-1}$. We can write (using $p = p_\text{d} + e$):

$$w = 0.622\frac{e}{p-e} \approx 0.622\frac{e}{p}\left(1+\frac{e}{p}\right) = q\left(1+\frac{q}{0.622}\right) = 1.016\,q. \qquad (5.38)$$

Note that the pressure of 800 hPa really did not matter. □

Example 5.5 Moistening the tropical boundary layer (temperature 300 K). Suppose the air above the sea surface is still (i.e., ignore advection) and absolutely dry. Suppose evaporation takes place steadily at a rate of 1 m yr^{-1}. How long does it take the boundary layer to come to 80% humidity?
Answer: The evaporation rate is 1 m yr^{-1}. Then

$$\frac{d\mathcal{M}}{dt} = \rho_\text{liq}V/At = 10^3\,\text{kg yr}^{-1}\,\text{m}^{-2}. \qquad (5.39)$$

At 300 K and 80% RH, the amount of water vapor above 1 m^2 in the boundary layer (1.5 km) is 32 kg. Hence,

$$t = \frac{32\,\text{kg}}{10^3\,\text{kg yr}^{-1}} = 0.032\,\text{yr} = 12\,\text{days}. \qquad (5.40)$$

□

Example 5.6 Can it saturate in time? Suppose a dry parcel is introduced to the tropical boundary layer at 30 °N. It flows along the surface in the trade winds until it reaches the Equator. Does it have time to saturate?
Answer: Let the trades flow toward the southwest at 10 m s^{-1}. The meridional distance to the Equator is 30×100 km $= 3 \times 10^6$ m. Along a diagonal at 45° this is enhanced to 4.24×10^6 m. The time required for this passage is approximately 4×10^6 s $= 46$ days. According to the previous example this appears to be sufficient time for the parcel to saturate. □

Example 5.7 The temperature is 20 °C and the vapor pressure is 10.0 Pa. What is the dew point temperature?
Answer: Solve (5.37) for T_D:

$$T_\text{D} = \left(\frac{1}{273.2} - \frac{R_\text{w}}{L}\ln\left(\frac{e}{6.11\,\text{hPa}}\right)\right)^{-1} = 280.2\,\text{K}. \qquad (5.41)$$

□

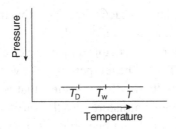

Figure 5.9 Relative positions of the temperature T, dew point T_D, and wet-bulb temperature T_w of an air parcel.

5.9 Wet-bulb temperature, LCL

The thermodynamic state of an air parcel is determined by its temperature, pressure and mixing ratio. However, in some applications it is not convenient to use the mixing ratio. For example, the mixing ratio is not easily measured directly. There is another indicator of atmospheric moisture called the wet-bulb temperature (T_w) which can be measured more directly. The wet-bulb temperature lies between the dew point temperature and the air temperature (see Figure 5.9).

Before introducing the wet-bulb temperature, let us turn again to the dew point temperature which is also a measure of moisture in the air. Recall that it is the temperature to which air must be cooled at constant pressure in order to become saturated with respect to a plane surface of water. As we perform the cooling we must keep the mixing ratio of the air fixed. At this temperature the actual mixing ratio becomes equal to the saturation mixing ratio:

$$w(T_D) = 0.622e_s(T_D)/p. \tag{5.42}$$

At temperatures higher than the dew point the air contains some moisture, but less than the saturation value. As the temperature is decreased to the dew point condensation occurs. Since at the dew point the mixing ratio is equal to the saturation mixing ratio, it is evident that the dew point temperature is always lower than or equal to the air temperature. If the two are close together, the relative humidity is high. If the dew point is far below the air temperature, the relative humidity is low. The dew point is a good indicator of human discomfort. When temperatures are high it is more comfortable with a low T_D rather than with a high dew point temperature because of the higher relative humidity. When the relative humidity is high the rate of evaporation from a moist surface is low (it is actually proportional to 100 minus the relative humidity in percent). Meteorologists often refer to the difference between temperature and dew point, which is called the *dew point depression*.

The frost point is defined similarly to the dew point. The frost point is the temperature to which air must be cooled at constant pressure in order to become saturated with respect to a plane surface of ice. Note that the mixing ratio is constant during a constant pressure cooling of a parcel.

The dew point temperature is difficult to measure directly. It is easy to measure the *wet-bulb temperature*. This is the temperature of a wet surface (nominally a wet cloth wrapped around the bulb of a thermometer) that is immersed in the ambient air. The wet surface evaporates moisture into the surrounding (typically less than saturated) air and in so doing the temperature of the wet surface is lowered (as perspiration into dry air cools the skin). The wet surface will come to an equilibrium temperature after a short time if the air near the wet surface is continually ventilated with the dryer ambient air. This equilibrium temperature of the wet cloth is called the wet-bulb temperature. If the surrounding air is fully saturated, there will be no net cooling since the rate of evaporation will just equal the rate of condensation onto the wet surface, leaving the wet-bulb temperature to be the same as the dew point temperature. For unsaturated air the wet-bulb temperature always falls between the dew point temperature and the dry-bulb temperature (the difference between the dry-bulb and wet-bulb temperatures is called the *wet-bulb depression*). Note that an evaporating cloud droplet or raindrop has a temperature at the wet-bulb temperature.

The wet-bulb temperature can be measured with a *sling psychrometer*, which consists of a thermometer with wet gauze covering its bulb. This hand-held device is swung around from a short chain to maintain the proximity of fresh ambient air at the wet surface. Without the swinging, stagnant saturated air would accumulate around the wet bulb and raise its temperature to an erroneous level. Water molecules leaving the wet surface diffuse away from the bulb through the thin boundary layer of air surrounding it. At the same time heat is being conducted from the warmer ambient air towards the cooler wet surface through the same thin boundary layer. Equilibrium is established between the enthalpy flux due to evaporation carried by out-flowing molecules and the in-flowing enthalpy flux. A formula can be derived for the relative humidity given the wet-bulb and dry-bulb temperatures (see *Wet-bulb derivation* in the box below). For practical use the relationship is commonly expressed in tables. It is interesting that the geometrical configuration of the wet cloth surrounding the bulb does not matter because those factors cancel. More discussion of the wet-bulb temperature can be found in Exercises 7.8 and 7.9 in Chapter 7.

The saturation mixing ratio depends on temperature and air pressure, thus it is a function of height. When a parcel is lifted adiabatically, its temperature and pressure both decrease. The temperature dependence of the parcel is linear with height (we will see in the next chapter that it is $10 \, \text{K} \, \text{km}^{-1}$). Hence, as the air rises 1 km the temperature will fall about 10 K. The saturation vapor pressure becomes half its surface value because of this decrease (remember the rule of thumb about the doubling of vapor pressure for every $10 \, °\text{C}$). The mixing ratio w_0 stays the same for this ascent, while the air pressure $p(z)$ and the vapor pressure $e(z)$ in the parcel

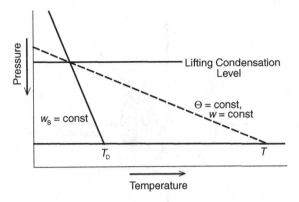

Figure 5.10 Illustration of the lifting condensation level (LCL).

each fall about 12% (using a scale height H of the atmosphere of about 8 km):

$$e(z) = \frac{p(z)w_0}{0.622}, \quad p(z) = p_0 e^{-z/H}, \quad w = w_0 = \text{constant}, \quad e^{-1/8} = 0.882. \quad (5.43)$$

So the vapor pressure goes down by $\approx 12\%$ while the saturation vapor pressure goes down by 50%. This shows that the saturation vapor pressure is decreasing much faster than the actual vapor pressure in the parcel. As this continues (the same percentages for each kilometer of ascent) the curves will cross and condensation will occur. The level at which an initially unsaturated parcel reaches its saturation level while being lifted adiabatically is called the *lifting condensation level* (LCL). If we know the temperature, pressure and mixing ratio of an air parcel, we can find the LCL, where condensation starts to occur along the ascent.

If we know the mixing ratio, we can easily find the dew point and wet-bulb temperature and vice versa. Figure 5.10 shows the LCL for a parcel at initial temperature T, pressure p, and dew point T_D. It is located at the intersection of the line with constant potential temperature (the unsaturated parcel ascends dry adiabatically) and the line of constant saturation mixing ratio starting at the dew point temperature (this is because the mixing ratio is fixed at its initial value during this noncondensing part of the ascent). The physical significance of these parameters will be more clear when we start to work with thermodynamic diagrams in Chapter 7.

Wet-bulb derivation In the preceding we indicated that the geometrical factors cancel in the wet-bulb depression. To show this we can derive the relationship between the relative humidity and the wet-bulb depression. For simplicity we take the bulb to be a sphere of radius R_0, but in the end it does not matter what the geometrical

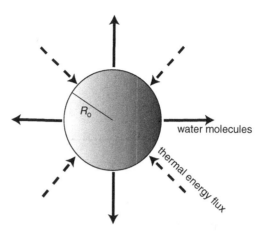

Figure 5.11 Water vapor molecules diffuse away from the wet surface as heat is conducted towards the wet surface. When these fluxes match, the temperature at the wet surface will be the wet-bulb temperature.

shape is. There are two fluxes from the sphere that have to be computed: (1) the flux of water vapor molecules leaving the sphere due to evaporation toward less moist air far away; (2) the heat conducted to the sphere because of the warmer air far away (Figure 5.11).

The water vapor molecules diffuse away with a flux $F_w(r)$ where r is the distance from the center of the sphere, $r \geq R_0$. The flux at any value of r passing through a surrounding sphere is

$$F_w(r) = -D4\pi r^2 \frac{dn_w}{dr} \tag{5.44}$$

where D is the diffusion coefficient for this process (known from tables), $n_w(r)$ is the number density of vapor molecules at radius r. This flux must be independent of r since otherwise there would be a source of flux other than the wet bulb's surface. We can then set $F_w(r) = F_0 = \text{constant}$. This allows us to integrate each side from R_0 to ∞ after dividing through by $4\pi r^2 D$. This produces:

$$n_\infty - n_{R_0} = -\frac{F_0}{4\pi R_0 D}. \tag{5.45}$$

We can recognize that

$$n_{R_0} = n_{\text{sat}}(T_w) \tag{5.46}$$

where $n_{\text{sat}}(T_w)$ is the saturation number density for $T = T_w$. If we divide through by $n_{\text{sat}}(T_D)$ we have

$$\text{RH} - \frac{n_{R_0}}{n_{\text{sat}}(T_D)} = \frac{-F_0}{4\pi R_0 D n_{\text{sat}}(T_D)} \tag{5.47}$$

where RH $= n_\infty/n_{sat}(T_D)$ is the relative humidity away from the wet bulb. So far we
do not know the value of F_0 in the above formula. Another measurement is required.
We must now turn to condition (2), the heat conduction. Heat flows from infinity
toward the wet bulb by heat conduction. The flux of heat crossing a spherical surface
of radius r is

$$H(r) = 4\pi r^2 \kappa_H \frac{dT}{dr} = H_0 = \text{constant} \tag{5.48}$$

where κ_H is the (available from tables) thermal conductivity coefficient and the same
argument is used as above for the constancy of the flux passing through spheres of
different radii.

As above we can integrate after dividing through by the factor $4\pi r^2 \kappa_H$. We get:

$$T_D - T_w = \frac{H_0}{4\pi R_0 \kappa_H}. \tag{5.49}$$

There is one other condition, that the flux of heat has to be related to the flux of
vapor molecules. Each molecule of water vapor leaving to infinity cools the wet bulb
by $\ell = \overline{L}/N_A$ where $\overline{L}$ is the enthalpy of vaporization per mole of water vapor and N_A
is Avogadro's number. This says that

$$H_0 = \ell F_0. \tag{5.50}$$

Now we can substitute for H_0 and divide the two equations above to obtain:

$$\text{RH} = \frac{n_{sat}(T_w)}{n_{sat}(T_D)} - \frac{\kappa_H}{\ell D} \frac{(T_D - T_w)}{n_{sat}(T_D)}. \tag{5.51}$$

Using the psychrometer we measure both T_D and T_w. We can then calculate $n_{sat}(T_D)$
and $n_{sat}(T_w)$ from the Clausius–Clapeyron relation. The only unknown above is RH,
which can now be calculated. The computation is tedious, hence, the tables. Note that
the geometric factors $4\pi R_0$ cancelled out. It can be shown that no matter what the
geometrical shape of the web bulb, these geometrical quantities will cancel out and
the wet-bulb temperature is independent of the shape of the bulb. So why swing
the psychrometer around? The reason is to have fresh environmental air within a
millimeter or two of the wet bulb to insure that the moist air at infinity is
representative and not contaminated by the wet bulb's evaporation.

5.10 Equilibrium vapor pressure over a curved surface

So far our discussion of the saturation vapor pressure has been restricted to that
over a plane surface of water. However, in atmospheric physics we also encounter
situations where the surface is curved. This is the case, for example, in cloud droplet
formation. Cloud droplets have approximately spherical shape, which means that
growth of a droplet implies an increase of the surface area of the drop. Increasing

surface area of a liquid requires work (consider blowing a soap bubble). Growth of a droplet then requires consideration of *surface tension*. Thus to find the equilibrium vapor pressure over a droplet we have to include surface tension in the energy balance. The energy required for an increase of surface area dA is $\sigma \, dA$, where σ is the surface tension (surface energy per unit area), in $J \, m^{-2}$ (the value for water is $0.0761 \, J \, m^{-2}$ at $0 \, ^\circ C$). The energy required to create a spherical drop[4] of radius a is $\sigma 4\pi a^2$.

Consider the formation of a cloud droplet from pure water vapor (no aerosols or other impurities present). Such a process is called *homogeneous nucleation* as opposed to *heterogeneous nucleation*, when small aerosol particles take part in the droplet formation. Suppose that initially, at $t = t_1$, we have water vapor of mass $\mathcal{M}$ at partial pressure e and temperature T. The Gibbs energy of the system, $G_{initial}$, is

$$G_{initial} = \mathcal{M} g_v \tag{5.52}$$

where g_v is the specific (per unit mass) Gibbs energy of the water vapor. It depends on the vapor pressure and the temperature. Suppose that at some later time, $t = t_2$, an embryonic droplet starts to form. It grows by occasional sticking collisions by water vapor molecules and at some moment develops a radius a and mass $\mathcal{M}_w$. The spherical surface area of the droplet is $A = 4\pi a^2$. The total mass is conserved, so if $\mathcal{M}_v$ is the mass of water vapor remaining after condensation, then

$$\mathcal{M}_v + \mathcal{M}_w = \mathcal{M}. \tag{5.53}$$

The total Gibbs energy of the system at time t_2, G_{final}, is

$$G_{final} = g_v \mathcal{M}_v + g_w \mathcal{M}_w + \sigma A. \tag{5.54}$$

The first term on the right-hand side of (5.54) is the Gibbs energy of the water vapor, the second is the Gibbs energy of the liquid, and the last term is due to surface tension.

The change of the Gibbs energy due to the droplet formation can be found by subtracting (5.52) from (5.54) and taking into account (5.53):

$$G_{final} - G_{initial} = (g_w - g_v)\mathcal{M}_w + \sigma A. \tag{5.55}$$

The next step is to find the difference $g_w - g_v$. We know that at a constant temperature (in our case the temperature is fixed) the change in the Gibbs energy is $dg = v \, dp$,

---

[4] The example of a soap bubble is helpful. Soapy water has a higher surface tension coefficient than pure water so the surface tension is very important. Also one must remember that the bubble has twice the surface area because there is an inside surface as well as an outside surface, in contrast to the water droplet.

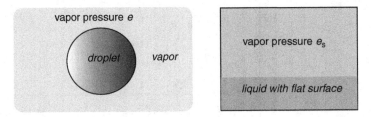

Figure 5.12 Schematic diagram illustrating the notation for equilibrium vapor pressure over a droplet, e, and that over a flat surface, e_s.

where v is the specific volume ($v = $ (density)$^{-1}$). Then, for the vapor we have $dg_v = v_v\, de$, for the liquid $dg_w = v_w\, de$, and the difference is equal to

$$d(g_v - g_w) = (v_v - v_w)de \approx v_v\, de \qquad (5.56)$$

since $v_v \gg v_w$. Substituting specific volume v_v from the Ideal Gas Law, we obtain

$$d(g_v - g_w) = R_w T \frac{de}{e} = R_w T d(\ln e) \qquad (5.57)$$

and, after the integration,

$$g_v - g_w = R_w T \ln e + \text{constant.} \qquad (5.58)$$

We can find the constant of integration by taking into account that in equilibrium (see Figure 5.12), along the phase boundary where $e = e_s(T)$, $g_w = g_v$ (see (5.7)). Then,

$$g_v - g_w = R_w T \ln \left(\frac{e}{e_s} \right). \qquad (5.59)$$

Substituting (5.59) into (5.55) gives

$$G_{\text{final}} - G_{\text{initial}} = -R_w T \ln \left(\frac{e}{e_s} \right) \mathcal{M}_w + \sigma A. \qquad (5.60)$$

The mass of a spherical water droplet with radius a and density ρ_w is $\mathcal{M}_w = \frac{4}{3}\pi \rho_w a^3$, and the surface area is $A = 4\pi a^2$. Then the change in the Gibbs energy due to the droplet formation, $\Delta G = G_{\text{final}} - G_{\text{initial}}$, is:

$$\Delta G = -\frac{4}{3}\pi a^3 \rho_w R_w T \ln \left(\frac{e}{e_s} \right) + 4\pi a^2 \sigma. \qquad (5.61)$$

From (5.61) we see that in the subsaturated air, when $e < e_s$, $\ln (e/e_s)$ is negative, and ΔG is always positive (see Figure 5.13). From Section 4.8.2 we know that equilibrium occurs when the minimum of the Gibbs energy is achieved; or, in other

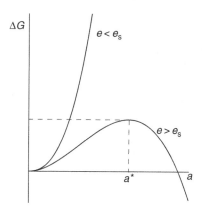

Figure 5.13 The change of Gibbs energy due to the formation of a water droplet as a function of the radius of the droplet.

words, the system tends spontaneously toward its equilibrium state by having the Gibbs energy diminish more and more until it reaches its minimum. We conclude that there are no favorable conditions for cloud droplet ($a > 0$) formation in unsaturated air.

The situation is different when the air is supersaturated, which means that $e > e_s$ and ln (e/e_s) is positive (see Figure 5.13). In this case ΔG increases as the radius a increases, then it reaches a maximum at some radius, and then decreases with the further increase of a. If the cloud droplet has a radius less than the critical radius a^*, it will disappear by evaporation. If, however, the cloud droplet reaches the critical radius a^*, then it will continue to grow. We can find an expression for the critical radius by equating the derivative $\partial G / \partial a$ to zero. The result is

$$a^* = \frac{2\sigma}{\rho_w R_w T \ln (e/e_s)} \tag{5.62}$$

which is called Kelvin's formula. It allows one to find the radius a^* of a droplet which is in equilibrium with air with the vapor pressure e. This equilibrium state is unstable. This becomes evident if we consider a slight growth of the droplet. This growth might be due to the condensation of water vapor in the vicinity of the droplet, which means that the relative humidity decreases just above the surface. Since the air in the vicinity of the droplet becomes drier, there is a diffusive flux of moist air toward the droplet, and the condensation process continues, leading to the further growth of the droplet. On the other hand, if the droplet evaporates slightly, the relative humidity just above the droplet surface increases, the water vapor starts to diffuse from the droplet, and the droplet will continue to evaporate to maintain the relative humidity corresponding to radius a^*.

We can rewrite (5.62) in order to determine the equilibrium vapor pressure e over a droplet with radius a,

$$e = e_s \exp\left(\frac{2\sigma}{\rho_w R_w Ta}\right) = e_s e^{b/a} \tag{5.63}$$

where the parameter b is defined by

$$b = \frac{2\sigma}{\rho_w R_w T}. \tag{5.64}$$

If we substitute $\sigma = 0.076$ J m^{-2}, $\rho_w = 1000$ kg m^{-3}, and $R_w = 461.5$ J kg^{-1} K^{-1}, we obtain

$$b = \frac{3.30 \times 10^{-7}}{T} \tag{5.65}$$

(in meters). At 273 K, $b = 1.21$ nm (typically $b \ll a$). Formula (5.63) shows that the equilibrium vapor pressure over a spherical droplet is not equal to the saturation vapor pressure as determined over a plane surface of water. This happens because of the surface tension. If the radius of the droplet goes to infinity, $a \to \infty$, which corresponds to a plane surface, we obtain the result for a flat surface $e = e_s$ (see Figure 5.12).

Example 5.8 Water droplets are in equilibrium with surrounding vapor at a temperature of 2 °C. Calculate the vapor pressure and relative humidity over the droplet with radius 0.008 µm.
Answer: The saturation vapor pressure at 2 °C is: $e_s = 2.497 \times 10^9 \exp(-5417/275.2)$ (hPa) $= 7.06$ hPa. From (5.63) we obtain $e = 7.06 \times \exp(3.3 \times 10^{-7}/(275.2 \times 0.008 \times 10^{-6})) = 8.2$ hPa. RH $= 116\%$. A supersaturation of 16% is required for the creation of a cloud droplet with radius 0.008 µm by homogeneous nucleation at 2 °C. □

To obtain the relative humidity, divide both sides of (5.63) by e_s:

$$\frac{e}{e_s} = e^{b/a}. \tag{5.66}$$

This relationship is illustrated in Figure 5.14 for a temperature of 5 ° C. As shown in the figure, the formation of a droplet with radius 0.01 µm requires supersaturation of 112%. At the same time, in real clouds the relative humidity rarely exceeds 101%. As we see from Figure 5.14, 1.0% supersaturation is required to form a droplet with radius larger than 0.1 µm. Large droplets in a cloud simply cannot form by random collisions of molecules on their surfaces. From the above discussion it follows that the process of homogeneous nucleation is unlikely. In nature cloud droplets form by heterogeneous nucleation on atmospheric aerosols. Consider then how the

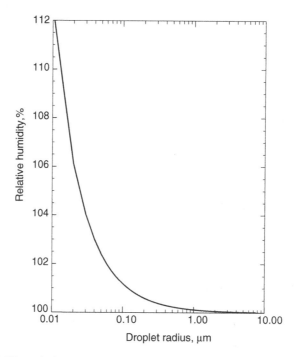

Figure 5.14 The relative humidity as a function of droplet radius. The curve is plotted for a temperature of 5 °C. This graph shows the relative humidity necessary to form a droplet of the size indicated on the abscissa.

equilibrium vapor pressure over the droplet changes if the droplet contains dissolved electrolytes (usually from aerosols, e.g. sea salt). We will consider *hygroscopic particles* (those that are soluble in water). The most common of these are sodium chloride (NaCl) and ammonium sulfate ($(NH_4)_2SO_4$) which dissolve when water condenses onto them. In this case a water droplet can be treated as a solution with the water as a solvent and the salt as a solute.

Chemistry refresher: Raoult's Law The ratio of the equilibrium vapor pressure over a solution, e', to the equilibrium vapor pressure over pure solvent, e, is equal to the mole fraction of the solvent in the solution, f, $e'/e = f$. The presence of salt in solution always lowers the vapor pressure.

Consider a cloud droplet that contains n_{salt} molecules of salt and n_w molecules of water per unit volume. If e' is the equilibrium vapor pressure over the solution (see Figure 5.15) then, according to Raoult's Law,

$$\frac{e'}{e} = \frac{n_w}{n_w + n_{salt}} = \frac{1}{1 + n_{salt}/n_w} \tag{5.67}$$

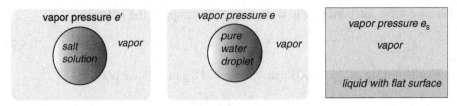

Figure 5.15 Illustration of the notation. Left: saturation vapor pressure e' over a solution droplet. Center: saturation vapor pressure e over a pure water droplet. Right: saturation vapor pressure over a flat surface e_s.

where e is the equilibrium vapor over a droplet composed of pure solvent. The second expression comes after dividing numerator and denominator by n_w. If we take into account that for a dilute solution $n_{salt} \ll n_w$ and expand the denominator in (5.67) in a geometric series,[5] we find (retaining only the linear term),

$$\frac{e'}{e} = 1 - \frac{n_{salt}}{n_w}. \tag{5.68}$$

In (5.68) the effect of dissociation of the ions is not yet taken into account. If the salt dissociates into i ions, say Na^+ and Cl^- giving $i = 2$, then the number of moles of solute individuals in the droplet, ν_{salt}, is

$$\nu_{salt} = i \frac{\mathcal{M}_{salt}}{M_{salt}} \tag{5.69}$$

where $\mathcal{M}_{salt}$ and M_{salt} are the mass and molecular weight of the salt respectively (be sure to use kg per mole here for the molecular weight). The degree of ionic dissociation $i \approx 2$ for sodium chloride and $i \approx 3$ for ammonium sulfate. The number of moles of water with molecular weight M_w in the mass $\mathcal{M}_w$ is

$$\nu_w = \frac{\mathcal{M}_w}{M_w} = \frac{4\pi a^3 \rho_w}{3 M_w} \tag{5.70}$$

where a is the radius of the droplet.[6] The ratio of moles is equal to the ratio of number densities:

$$\frac{n_{salt}}{n_w} = \frac{\nu_{salt}}{\nu_w} = \frac{3i \mathcal{M}_{salt} M_w}{4\pi a^3 \rho_w M_{salt}}. \tag{5.71}$$

After substituting (5.71) into (5.68) we obtain

$$\frac{e'}{e} = 1 - \frac{c}{a^3} \tag{5.72}$$

[5] The geometric series for $\frac{1}{1+\epsilon} = 1 - \epsilon + \epsilon^2 - \cdots$, where $|\epsilon| < 1$.
[6] We assume that dissolving the salt particle does not change the volume of the droplet.

where we have introduced a new parameter to simplify the notation

$$c = 3i\mathcal{M}_{salt}M_w/4\pi\rho_w M_{salt}. \tag{5.73}$$

After substituting $\rho_w = 1000\,\text{kg m}^{-3}$, $M_w = 18\,\text{g mol}^{-1}$ we obtain

$$c \approx 4.3 \times 10^{-6} i \frac{\mathcal{M}_{salt}}{M_{salt}}\ (\text{m}^3). \tag{5.74}$$

Combining Kelvin's formula (5.66) and (5.72) we obtain the formula for equilibrium vapor pressure of a solution droplet:

$$\frac{e'}{e_s} = \left(1 - \frac{c}{a^3}\right)e^{b/a}. \tag{5.75}$$

To better understand this formula consider a limiting case. We again let the radius of the droplet go to infinity, $a \rightarrow \infty$. From (5.75) we obtain the known result: the equilibrium vapor pressure over the plane surface of water ($a = \infty$) is equal to the saturation vapor pressure, $e' = e_s$.

Consider now a droplet with a radius $a \gg b$. Then one can expand the exponential function in (5.75) in a Taylor series [7] and get

$$\frac{e'}{e_s} = 1 + b/a - c/a^3. \tag{5.76}$$

The second term on the right-hand side of (5.76) is responsible for the surface tension, the third term is due to the presence of the salt in the droplet. A graphical illustration of formula (5.76) is given in Figure 5.16 for $10^{-19}\,\text{kg}$ of sodium chloride. The curve showing the dependence of relative humidity on radius of the solution droplet is called the Köhler curve. The dashed curve corresponds to homogeneous nucleation (5.63) with no salt present. These two curves differ considerably for small values of the radius of the solution droplet. The most important result is that with an embedded soluble salt nucleus in the droplet a much lower supersaturation is required for the droplet to be in equilibrium with its environment than in the case of a pure water droplet of the same size. For example, with supersaturation of just 0.1% a droplet with a radius slightly larger than 0.1 μm can be formed. With a very small radius the droplet can be in equilibrium with the surrounding air even with a relative humidity less than 100%. This is possible only because of the presence of hygroscopic particles. With relative humidity 90% a solution droplet with radius 0.05 μm can form if there are $10^{-19}\,\text{kg}$ of sodium chloride (not shown in the graph).

For a small droplet radius, the Köhler curve increases monotonically until it reaches a maximum at radius $a = a'$, and then it decreases monotonically. Consider

---

[7] $e^x \approx 1 + x$, for $|x| \ll 1$.

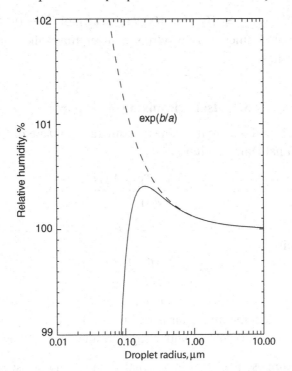

Figure 5.16 Relative humidity with which a solution droplet formed on 10^{-19} kg of NaCl is in equilibrium with water vapor as a function of the droplet radius. The temperature is 5 °C.

a droplet with a radius less than a'. If the relative humidity increases, such a particle grows to adjust to the new equilibrium conditions. The equilibrium is stable, since slight fluctuations in condensation or evaporation do not lead to further growing or shrinking of the droplet. If the droplet grows slightly by condensation, the relative humidity increases, the diffusive flux is directed away from the droplet and the droplet evaporates to return to equilibrium. With slight evaporation, the relative humidity in the vicinity decreases, which causes diffusive flux toward the droplet, and the droplet again returns to equilibrium with the surrounding air. Such droplets with radius less than a' are called haze particles.

The situation is different if the droplet reaches the radius a'. Now the equilibrium is unstable. With any further growth of the droplet the relative humidity decreases and the droplet continues to grow. This is one mechanism for cloud droplet formation.

Example 5.9 How big is a salt particle whose mass is 10^{-19} kg? Using the density of 2165 kg m^{-3}, we can calculate that a spherical particle of this mass would have a radius of 0.022 μm. How many molecules are in this particle? The number of moles is $\mathcal{M}/M_{NaCl} = 10^{-19} \times 1000/58.44\,\mathrm{kg\,mol^{-1}} = 0.017 \times 10^{-16}\,\mathrm{mol}$. The number of molecules is Avogadro's number times the number of moles: 1.03×10^6 molecules. $\qquad\qquad\qquad\qquad\qquad\qquad\qquad\qquad\qquad\qquad\qquad\qquad$ □

5.11 Isobaric mixing of air parcels

When two parcels of dry air at the same pressure (altitude) are brought into contact and mixed, the final temperature is

$$T_f = \frac{\mathcal{M}_1 T_1 + \mathcal{M}_2 T_2}{\mathcal{M}_1 + \mathcal{M}_2}. \qquad (5.77)$$

If the mixing ratios of the parcels are w_1 and w_2, we can arrive at the same kind of linear relationship

$$w_f = \frac{\mathcal{M}_1 w_1 + \mathcal{M}_2 w_2}{\mathcal{M}_1 + \mathcal{M}_2}. \qquad (5.78)$$

It is possible that although the combinations (w_1, T_1) and (w_2, T_2) are neither one individually saturated, the mixed air parcel (w_f, T_f) is saturated. Two clear moist air parcels can come into contact with a foggy result.

Example 5.10 Suppose two parcels of equal volumes and pressures, but differing temperatures (273 K and 293 K) come into contact near the ground ($p = 1000\,\mathrm{hPa}$). Let each have relative humidity 90%. We can use

$$p = \frac{\mathcal{M}_1}{V} R T_1 = \frac{\mathcal{M}_2}{V} R T_2. \qquad (5.79)$$

Then

$$\frac{\mathcal{M}_1}{\mathcal{M}_2} = \frac{T_2}{T_1}. \qquad (5.80)$$

The final temperatures are:

$$T_f = \frac{2 T_2 T_1}{T_1 + T_2} = 283\,\mathrm{K}. \qquad (5.81)$$

We can compute the initial mixing ratios:

$$w_1 = 0.622 \times 0.9 \times 6.11/1000 = 0.0034, \qquad w_2 = 0.0131. \qquad (5.82)$$

The final mixing ratio is

$$w_f = 0.0083. \qquad (5.83)$$

But the saturation mixing ratio at T_f is

$$2.497 \times 10^8\, e^{-5417/283} \times 0.622/1000 = 0.0076. \qquad (5.84)$$

We find that the final mixing ratio is above the saturation value for the final temperature, hence we have supercooled water vapor and therefore fog.

The final temperature will actually be slightly above T_f because some heating will occur during the condensation. □

Notes

More advanced treatments of water and air can be found in Bohren and Albrecht (1998), Irebarne and Godson (1981), and Curry and Webster (1999). Discussions of cloud drops, etc., can be found in Fleagle and Businger (1980), Rogers and Yau (1989), Houze (1993) and Emanuel (1994).

Notation and abbreviations for Chapter 5

a	droplet radius (m)
a^*	critical droplet radius (m)
e, e_s	vapor pressure, saturation vapor pressure (Pa)
e'	vapor pressure over a solution (Pa)
$\epsilon = M_w/M_d = 0.622$	(dimensionless)
g_l, g_g	specific Gibbs energy for liquid, gas (J kg^{-1})
g_v, g_w	specific Gibbs energy for vapor, liquid water (J kg^{-1})
G	Gibbs energy (J)
h	specific enthalpy (J kg^{-1})
$H(r)$	flux of heat crossing the surface of a sphere (J s^{-1})
κ_H	thermal conductivity coefficient (J m K^{-1} s^{-1})
$L = \Delta H_{vap}$	the enthalpy (latent heat) of evaporation (J kg^{-1})
LCL	lifting condensation level
M_v, M_d, M_e	gram molecular weight of vapor, dry air and effective (g mol^{-1})
$\mathcal{M}_l, \mathcal{M}_g$	bulk mass of liquid, gas (kg)
n_s	number density of vapor molecules at saturation (molecules m^{-3})
n_{sat}	number density of vapor molecules at saturation (molecules m^{-3})
n_w	number density of water molecules in vapor (molecules m^{-3})
n_0	number density (molecules m^{-3})
N_A	Avogadro's number (molecules mol^{-1})
q	specific humidity (kg water vapor/kg of moist air)
r	relative humidity
R_w	the gas constant for water vapor (Table 1.1)

R_{eff} effective gas constant for a mixture of species
          $(\text{J kg}^{-1}\,\text{K}^{-1})$
R^* universal gas constant $(\text{J mol}^{-1}\,\text{K}^{-1})$
s_1, s_g specific entropy for liquid, gas $(\text{J K}^{-1}\,\text{kg}^{-1}))$
S entropy (J K^{-1})
σ surface tension (J m^{-2})
T Kelvin temperature (K)
T_D dew point temperature (K)
T_v virtual temperature (K)
T_w wet-bulb temperature (K)
θ potential temperature (K)
$\bar{v}$ mean molecular speed (m s^{-1})
v_l, v_g specific volume for liquid, gas $(\text{m}^3\,\text{kg}^{-1})$
w mixing ratio (kg water vapor per kg of dry air)
w_s saturation mixing ratio (kg water vapor per kg of dry air)

Problems

5.1 A kilogram of water is vaporized at $0\,^\circ$C and at 1000 hPa atmospheric pressure. (a) Calculate the change in enthalpy of the water substance in the transition and (b) the change of entropy for the process.

5.2 What is the virtual temperature for a kilogram of air at $T = 283$ K, relative humidity 50% at 1000 hPa?

5.3 The normal temperature of human blood is $37.2\,^\circ$C. If a person is lifted in a balloon the air pressure decreases. There will be a pressure (altitude) where the blood begins to boil. What is that pressure in hPa? At about how many meters is that above sea level?

5.4 It is a muggy night at the old ball park. The temperature is $30\,^\circ$C and the humidity is 85%. What is the change in density (%) of the air from a dry night at the same temperature and pressure (1000 hPa)?

5.5 Assume the atmosphere is isothermal at 303 K (very tropical), which gives a scale height of $H = 8.87$ km. The surface humidity is 80%. The vapor is distributed uniformly in the first 1.5 km, and the air is dry above that. For simplicity take the air pressure to be uniform in this lowest 1.5 km. How much water (in vapor form) lies above a given square meter of surface (kg m^{-2})? Express the result in mm equivalent of liquid water. Compare to the situation when the temperature is 273 K.

5.6 A system consists of dry air mixed with water vapor at a temperature of $20\,^\circ$C. The pressure of the mixture is 990 hPa. The relative humidity is 50%.

   (a) What is the saturation vapor pressure?
   (b) What is the partial pressure of the water vapor?
   (c) What is the density of the mixture. Compare it with the density of dry air at the same T and p.

(d) What are w and w_s?

(e) If the system (parcel) is lifted adiabatically to 500 hPa, which is conserved e or w?

5.7 Calculate the equilibrium vapor pressure over spherical droplets with radii 0.01, 0.1, 1, 10 μm at temperature 273 K. Plot the relative humidity (with respect to a flat water surface) as a function of radius.

5.8 What supersaturation is needed for the droplets with radius 0.5 μm to be in the equilibrium with water vapor at temperature of 10 °C?

5.9 An ammonium sulfate $((NH_4)_2SO_4)$ particle of mass 10^{-20} kg of radius 0.07 μm is present in the air at temperature 0 °C. Find the relative humidity necessary for heterogeneous nucleation.

5.10 Find the expression for the critical value (maximum of the Köhler curve) of the droplet radius and relative humidity. Calculate these values for a droplet containing 10^{-16} kg of NaCl at 0 °C.

6

Profiles of the atmosphere

The properties of the atmosphere that vary with altitude include the pressure, temperature and the composition of constituents such as water vapor. This chapter provides some insight into these dependencies with simple derivations that hold under idealized conditions. The stage will be set for the following chapter which provides methods for analyzing the conditions at the time of observation.

6.1 Pressure versus height

Atmospheric pressure drops off dramatically with height above the surface. This is indicated by the graph in Figure 6.1 which shows the dependence of pressure on altitude for the US Standard Atmosphere.[1]

By balancing the vertical components of force on a slab of air at an arbitrary height z it is possible to derive a formula for the average pressure as a function of height, $p(z)$. Consider a column of air with cross-section $1\,m^2$. In the column

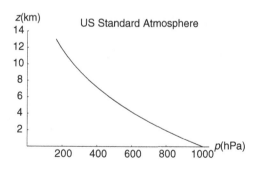

Figure 6.1 Dependence of pressure p (hPa) on altitude z (km) for the US Standard Atmosphere.

---

[1] The US Standard Atmosphere is a model of the atmospheric profile of various properties. It is used primarily in aviation and satellite drag calculations. It attempts to give a global average of conditions. More can be learned about it on the internet.

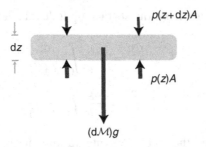

Figure 6.2 Diagram of a column of air of cross-section area A with a slab of thickness dz at height z. The pressure and gravitational forces on the slab are indicated.

we picture a thin horizontal slab of air whose bottom surface is located at height z above sea level and whose thickness is dz (see Figure 6.2). The mass of material in the slab is (density times volume): $d\mathcal{M} = \rho A dz$, where $A = 1\,\mathrm{m}^2$ is the horizontal cross-sectional area of the slab. The weight of the slab of gas is $(d\mathcal{M})g = (\rho A dz)g$.

Beneath the slab is a pressure force pushing upwards: $p(z)A$. Above the slab is a pressure force pressing downwards,

$$p(z + dz)A \approx \left(p(z) + \frac{dp}{dz}\,dz \right) A. \tag{6.1}$$

Equating the net pressure force on the slab to the gravitational force,

$$\frac{dp}{dz}\,dz\,A = -\rho g\,dz\,A. \tag{6.2}$$

After cancellations we obtain the *hydrostatic equation*:

$$\boxed{\frac{dp}{dz} = -\rho g}\quad \text{[hydrostatic equation]}. \tag{6.3}$$

We can use the Ideal Gas Law to write:

$$\frac{dp}{dz} = -\frac{p(z)g}{RT(z)} \tag{6.4}$$

where we indicate explicitly that both temperature and pressure are functions of altitude z. In the last step we used the ideal gas equation of state. The hydrostatic equation has many uses, but it is particularly useful if the dependence of T on z is known. This may often be the case, at least approximately. If it is true we can write:

$$\frac{dp}{p} = -\frac{g\,dz}{RT(z)}. \tag{6.5}$$

Then integrating each side from the surface up to level z leads to:

$$\int_{p_0}^{p(z)} \frac{dp}{p} = -\frac{g}{R} \int_0^z \frac{dz'}{T(z')} \tag{6.6}$$

$$\ln \frac{p}{p_0} = -\frac{g}{R} \int_0^z \frac{dz'}{T(z')} \tag{6.7}$$

where we have indicated the "dummy" integration variable by z' to distinguish it from the upper limit of the integral. Finally,

$$p(z) = p_0 \exp \left(-\frac{g}{R} \int_0^z \frac{dz'}{T(z')} \right). \tag{6.8}$$

If the integrals can be performed, we have an analytical expression for $p(z)$. Even if the z dependence of T is known only graphically or in tabular form, the integral can be performed numerically.

Example 6.1 In Example 2.16 we found that a ball of mass m bouncing elastically on the floor gives an average force on the floor of mg. We can also derive the hydrostatic equation for a ball bouncing on the floor, but reflecting back elastically on a ceiling only a short distance above. Let the ball leave the floor with vertical velocity v_0 and when it gets to the ceiling h, its velocity will be v_1. The rate of momentum transfer to the ceiling is $2mv_1/T$, where T is the time for a round trip ceiling to floor and back. We can show that $T = 4h/(v_0+v_1)$. The average pressures exerted by the reflecting ball at the ceiling p_h and at the floor p_0 are

$$p_h = \frac{mv_1(v_0 + v_1)}{2hA}, \quad p_0 = \frac{mv_0(v_0 + v_1)}{2hA}, \tag{6.9}$$

where A is the area on the surface of the floor or ceiling, and the difference in the pressures is

$$p_0 - p_h = \frac{m}{2hA}(v_0 - v_1)(v_0 + v_1) = \frac{m}{2hA}(v_0^2 - v_1^2) = \frac{m}{2hA}2gh. \tag{6.10}$$

We then have

$$p_0 - p_h = \frac{mgh}{Ah} = \frac{mgh}{\text{volume}} = \rho gh. \tag{6.11}$$

Finally, $\Delta p/\Delta z = -\rho g$. Of course, we are to picture a large number of balls (molecules) bouncing up and down so as to make a steady pressure. □

Example 6.2: constant density case Consider the case of constant density as a function of height. This profile is more like the ocean than the atmosphere. Taking

$z = 0$ to be at the bottom of the fluid, we can integrate the hydrostatic equation directly from 0 to the top of the fluid H:

$$\int_{p_0}^{0} dp = -\rho_0 g \int_{0}^{H} dz \qquad (6.12)$$

$$p_0 = \rho_0 g H. \qquad (6.13)$$

Integrating from 0 to z we find that

$$p(z) = \rho_0 g H \left(1 - \frac{z}{H} \right). \qquad (6.14)$$

The pressure is zero at the top of the fluid and it increases linearly with depth below the surface until the bottom ($z = 0$) is reached. □

Example 6.3: pressure in the ocean What is the pressure at 5 km below the surface in the ocean?
Answer: The density of water is $10^3 \, \text{kg m}^{-3}$. Now $\rho g h = 5 \times 10^7 \, \text{Pa} \approx 500 \, \text{atm}$. For reference, it might be noted that 1 atm of pressure is equivalent to that at about 10 m depth of water. □

Example 6.4: isothermal atmosphere Take the temperature to be constant ($T = T_0$) with respect to z. Then we can integrate:

$$\frac{dp}{p} = -\frac{g}{RT_0} dz \qquad (6.15)$$

and after integration from $z = 0$ to z:

$$p(z) = p_0 e^{-z/H} \qquad (6.16)$$

where p_0 is the pressure at $z = 0$ and the *scale height H* is given by

$$H = \frac{RT_0}{g}. \qquad (6.17)$$

Note the straightforward physical interpretation: large temperatures lead to large scale heights (swelling); large g leads to smaller scale heights. Moreover, larger R (smaller molecular weight) leads to larger H. □

It turns out that taking the scale height H to be a constant is a rather good approximation to the pressure dependence on altitude, as can be seen in Figure 6.3 wherein a calculated profile (dashed) is superimposed on the US Standard Atmosphere values. In the calculation a value of $H = 7.89 \, \text{km}$ was used along with $p_{z=0} = 1013 \, \text{hPa}$. This model fit to the empirical result is very good considering that the temperature is very altitude dependent for the US Standard Atmosphere as shown in Figure 6.4.

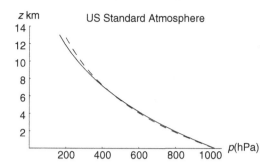

Figure 6.3 The pressure versus altitude curve with a computed profile (dashed) superimposed. The calculation was conducted with the constant value of $H = 7.89$ km and a value of $p_{z=0} = 1013$ hPa (to agree with the Standard Atmosphere at the surface).

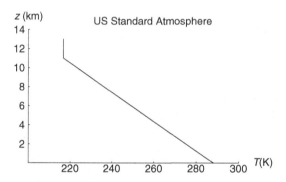

Figure 6.4 The temperature versus altitude curve for the US Standard Atmosphere. The altitude of the discontinuity in slope at about 11 km is called the *tropopause*.

6.2 Slope of the dry adiabat

Consider next the dry adiabat. This is the temperature dependence of a parcel of dry air as it is displaced upwards [2] under adiabatic conditions (no heating due to the temperature differential with the environment). For a small vertical displacement of the parcel we can write the change in enthalpy for the parcel (treated as a thermodynamic system undergoing a reversible transformation):

$$dH = \mathcal{M}c_p\, dT = V\, dp$$

$$= \mathcal{M}\frac{RT}{p}\, dp. \tag{6.18}$$

---

[2] Some external means must cause the displacement such as buoyancy or the rise of air due to upslope wind.

Dividing each side by $\mathcal{M}\,dz$,

$$c_p \frac{dT}{dz} = \frac{RT}{p}\frac{dp}{dz}, \tag{6.19}$$

and using the hydrostatic equation ($dp/dz = -\rho g$) and $\rho = p/RT$ we obtain:

$$c_p \frac{dT}{dz} = \frac{RT}{p}(-\rho g) = -g. \tag{6.20}$$

Finally,

$$\left.\frac{dT}{dz}\right|_{\text{dry adiabat}} = -\frac{g}{c_p} \quad \text{[dry adiabatic lapse rate].} \tag{6.21}$$

This very simple and elegant result does not depend on the actual temperature profile of the atmosphere. We can evaluate this formula to find:

$$\boxed{-\left.\frac{dT}{dz}\right|_{\text{dry adiabat}} = \Gamma_d = 10\,\text{K}\,\text{km}^{-1}} \tag{6.22}$$

Note that the adiabatic lapse rate Γ_d is defined to be a positive number. Figure 6.5 shows the temperature and size of a parcel being lifted adiabatically from the surface to 10 km.

Example 6.5: dry adiabatic atmosphere Consider the pressure profile of an atmosphere whose vertical dependence of temperature is that of a dry adiabat. This

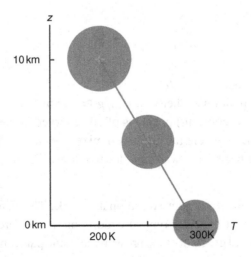

Figure 6.5 Illustration of the size of a parcel as it is lifted adiabatically. The volume of the spherical parcel was calculated from (3.43).

is an atmosphere that is thoroughly mixed in the vertical dimension. (For example, the turbulent boundary layer which occupies the lowest 1 to 2 km of the air column. We will show this later.) Its temperature falls off linearly as $dT/dz = -g/c_p$. For these conditions θ is the same throughout. Such an atmospheric profile is called *isentropic*. We have the hydrostatic equation

$$\frac{dp}{dz} = -\frac{p}{RT}g \tag{6.23}$$

and Poisson's equation

$$T = \theta \left(\frac{p}{p_0}\right)^\kappa . \tag{6.24}$$

We take θ to be a constant in Poisson's equation. Thus,

$$\frac{dp}{dz} = -\frac{pg\, p_0^\kappa}{\theta R\, p^\kappa} \tag{6.25}$$

$$\frac{p^{\kappa-1}}{p_0^\kappa} dp = -\frac{g}{\theta R} dz. \tag{6.26}$$

Integrating from $(p = p_0, z = 0)$ to $(p = p, z = z)$:

$$\left(\frac{p}{p_0}\right)^\kappa = 1 - \frac{\kappa gz}{\theta R} \tag{6.27}$$

and finally,

$$p(z) = p_0 \left(1 - \frac{gz}{\theta c_p}\right)^{1/\kappa} \tag{6.28}$$

where in the last we used $\kappa = R/c_p$.

Note that the formula fails when $z > \theta c_p/g \approx 30\,\text{km}$.

The isentropic ($\theta = \text{constant}$) profile is often observed in the daytime boundary layer and up to the LCL where the air is well mixed vertically. Above this level the lapse rate is smaller because of warming due to condensation of water vapor in the parcel into droplets.   □

When does the hydrostatic approximation not work? This can happen in unusual circumstances, but first consider typical conditions. If the force of gravity is not balanced by the vertical gradient of the pressure field, the parcel must be accelerating vertically. Suppose the imbalance is 1% or an acceleration of $\approx 0.1\,\text{m s}^{-2}$. After only 10 s a parcel starting at rest would have a vertical velocity of $1\,\text{m s}^{-1}$, a very rare occurrence except in a thunderstorm. Vertical velocities are typically of the order of $0.01 \, \mathrm{m \, s^{-1}}$, which suggests that large imbalances are very rare. Arguments can also be constructed from three-dimensional considerations that at synoptic scales (scales that match the typical observing stations on a weather map, a few hundred kilometers) one finds that horizontal motions are typically on the order of 1 to $10 \, \mathrm{m \, s^{-1}}$ and vertical motions are of the order of centimeters per second. These arguments can be found in the first few chapters of most dynamics books.

6.3 Geopotential height and thickness

The mechanical potential energy per unit mass of a parcel (called the *gravitational geopotential*) is $\Psi(z) = gz$ where z is its height above some reference level (typically sea level).[3] We can write for the change in potential from one level to another:

$$\Psi_{\text{above}} - \Psi_{\text{below}} = (z_{\text{above}} - z_{\text{below}})g. \tag{6.29}$$

This is the amount of work performed in lifting a 1 kg parcel from z_{below} to z_{above} (not counting buoyancy forces, just gravity). The geopotential can be turned around slightly to be considered a function of the pressure level of the parcel. So instead of $\Psi(z)$ we can think of $\Psi(p)$. This is just the gravitational potential energy per unit mass of a parcel at pressure level p. Now the change in gravitational potential energy in going from one pressure level to another is

$$\Psi_{\text{above}} - \Psi_{\text{below}} = (z_{p_{\text{above}}} - z_{p_{\text{below}}})g \tag{6.30}$$

where $z_{p_{\text{above}}}$ and $z_{p_{\text{below}}}$ are the elevations above the reference level for which the pressures are p_{above} and p_{below}. The *geopotential height, Z_p,* is defined to be the height in meters of the pressure level for a given value of the potential energy per unit mass:

$$Z_p = \Psi(p)/g. \tag{6.31}$$

The *geopotential height, $Z_{p_1}(x, y)$,* is the elevation of the surface for a given pressure $p = p_1$. The height of this constant pressure surface is a function of x and y (longitude and latitude) over the Earth's surface. All meteorologists are familiar with the 500 hPa height field, since it is so important in weather forecasting.

Example 6.6: height field of an isothermal atmosphere Suppose the temperature is everywhere T_0. What is the 500 hPa height field?

---

[3] If z were large enough we would have to take into account the z dependence of $g = g(z)$ (due to the weakening of the gravitational force with distance from the Earth's center) and use $\int^z g(z) \, dz$, but this is seldom important in studies of weather and climate of the troposphere.

Answer: For an isothermal atmosphere we have

$$p = p_0 e^{-z/H} \tag{6.32}$$

or after taking natural logs of each side,

$$z = H \ln \frac{p_0}{p}. \tag{6.33}$$

For a temperature of 300 K, the scale height $H = R_d T_0/g$ is 8786 m. If the reference pressure is fixed at 1000 hPa, using $\ln 2 = 0.693$, we have $Z_{500} = 6090$ m.

Note that because it is much colder at the poles, the 500 hPa height field is lower at the poles ($T \approx 250$ K) than at the Equator ($T_{Equator} \approx 300$ K), or about 5000 m versus 6000 m. Roughly speaking, the height field scales inversely to T, but remember this is for an isothermal atmosphere.

In our solution we also had to specify the surface pressure. If there is more atmospheric mass above one position on the Earth (as measured by surface pressure), this will lift the geopotential height field. So the height field is determined by the amount of mass above the reference surface (this sets the surface pressure) at a point and the thermal structure of the air above that point. Keep in mind that the surface pressure (proportional to the mass above the site) varies no more than a few percent from time to time. It falls as much as 8% or 9% in the eye of the most intense (hurricane strength) tropical storms. ☐

The vertically averaged temperature in a layer can be defined as

$$\boxed{\overline{T} = \frac{\int_{p_{below}}^{p_{above}} T \, d \ln p}{\int_{p_{below}}^{p_{above}} d \ln p}} \tag{6.34}$$

where $d \ln p = dp/p$. (Note that $dz \propto -dp/p$ at least locally according to the hydrostatic equation.) This justifies the use of $d \ln p$ as our integration increment. After applying the hydrostatic equation we find:

$$\overline{T} = \frac{g \int_{z_{above}}^{z_{below}} dz}{R \ln(p_{above}/p_{below})} = \frac{g}{R} \cdot \frac{z_{below} - z_{above}}{\ln(p_{above}/p_{below})}. \tag{6.35}$$

Finally,

$$z_{above} - z_{below} = \frac{R\overline{T}}{g} \ln \frac{p_{below}}{p_{above}} = \overline{H} \ln \frac{p_{below}}{p_{above}} \quad \text{[thickness]} \tag{6.36}$$

and for dry air $\overline{H} = 29.3\overline{T}$ is the (local) scale height. The quantity $\Delta z = z_{above} - z_{below}$ is called the *thickness* of the layer lying between the two pressure surfaces.

Clearly the thickness is a measure of the local vertically averaged temperature of the layer.

Example 6.7 How thick is the 500 to 600 hPa layer when the average temperature is 280 K?
Answer:

$$\Delta z \approx 1500 \, \text{m}. \tag{6.37}$$

□

Example 6.8 The 1000 hPa to 500 hPa thickness is often used in weather discussions to describe the average temperature in the lower part of the troposphere. What is the average temperature for a sea level temperature of 295 K and a lapse rate of 6 K km^{-1}?
Answer: Use $\overline{T} = (1/\Delta z) \int_0^{\Delta z} (T_0 - \Gamma z) \, dz$ and (6.36). Then the temperature $\overline{T} = 278$ K. The thickness is ≈ 5647 m.   □

It is important to notice that thickness is a measure of temperature. Cold layers are thin, warm layers thick, in direct proportion to the average Kelvin temperature in the layer.

6.4 Archimedes' Principle

In a fluid such as the atmosphere a low density parcel embedded in a denser environment will rise. Consider a fluid of uniform density. Figure 6.6 shows an arbitrary volume isolated inside a box of fluid of uniform density (picture a thin film enclosing a portion of the same uniform fluid). The fluid including the enclosed parcel is at rest. Therefore, all components of the forces acting on the enclosed interior portion (parcel) must sum vectorially to zero. We need only consider

$\rho_e V g$

Figure 6.6 Diagram of an irregular volume in a fluid. The weight of the mass in the volume is $\rho_e V g$, where ρ_e is the density of the fluid, V is the volume and g is the acceleration due to gravity. The pressure forces of surrounding fluid indicated by the arrows pointing inward exactly balance the downward pointing force due to gravity. But if a fluid of lower density, say ρ, is substituted in the same irregular volume, there will be a net upwards force on this mass.

the vertical components, since the horizontal components balance without any gravitational contribution. The weight of the interior volume indicated by the downward pointing vector (see Figure 6.6) has magnitude $\mathcal{M}g = \rho_e Vg$. The subscript e indicates the density of the uniform fluid (the environment). The pressure forces exerted by the surrounding fluid must add up vectorially to a vector pointing upwards with magnitude $\mathcal{M}g$. Now suppose the isolated volume is carved out and refilled with matter of a different density, say ρ. The force exerted on this volume (parcel) is upwards and of magnitude $\rho_e Vg$, but the weight of the parcel is only ρVg. There is thus a net upward force of

$$F = (\rho_e - \rho)Vg. \tag{6.38}$$

If the density of the parcel ρ is less than the environmental density, the force will be upwards. This is the *buoyancy force* on the parcel and the formula is called *Archimedes' Principle*. Often we want to know the force per unit mass or acceleration. We can use Newton's Second Law ($F = \mathcal{M}a$ where a is the vertical acceleration):

$$\boxed{a = \frac{\rho_e - \rho}{\rho} g} \quad \text{[Archimedes' Principle].} \tag{6.39}$$

We may use $p = p_e$ and for an ideal gas $\rho = p/RT$; hence

$$a = \left(\frac{1}{T_e} - \frac{1}{T}\right) T g \tag{6.40}$$

and simplifying

$$\boxed{a = \left(\frac{T - T_e}{T_e}\right) g} \quad \text{[acceleration by buoyancy].} \tag{6.41}$$

Note that if $T > T_e$, the force is upward; and if $T < T_e$, the force is downward. (Proof that warm air rises as we learned in elementary school!)

Example 6.9 Consider a thunderstorm shaft. Let the temperature inside the shaft near the surface be $5\,\text{K}$ warmer than that outside $\approx 300\,\text{K}$. Then the vertical acceleration of a parcel in the shaft is approximately $0.163\,\text{m s}^{-2}$. If the acceleration is constant the velocity after time t is $v = at$ and the altitude is $z = \frac{1}{2}at^2$. The parcel reaches $4\,\text{km}$ in about $3.7\,\text{min}$ at which time it has a vertical velocity of $36\,\text{m s}^{-1}$.

Of course, this analysis is very simplified because the effects of latent heat release are ignored. In addition it was assumed that the differential temperature between the interior of the parcel and the environment remained constant during the ascent. In ensuing sections we will see how these effects are taken into account.   □

6.5 Stability

A slab of air (parcel) sits at rest in an environment with a certain temperature profile. The forces are balanced on the parcel. What happens if the parcel is nudged upwards or downwards by a tiny amount? Do the forces in the new position balance or do they impart a restoring force or perhaps a repelling force? First consider a parcel at point A in Figure 6.7. Note that a straight line segment slanted upwards into the upper left quadrant direction passes through the point at A. This straight line segment is a dry adiabat passing through A. If a parcel is lifted adiabatically at A, it moves along this adiabat and it finds itself warmer than the environmental curve which lies to its left. In this case the parcel, being warmer than the environment, will experience a buoyant force upwards. Hence, a parcel at point A under a small perturbation upwards will experience a net buoyant force to continue going upward. If the same parcel is nudged downwards from A it will experience a downward buoyant force. We say the point A is *unstable* with respect to dry convection.

Next consider the point C. An analysis similar to the above shows that the point C is *stable* with respect to infinitesimal perturbations. The point B is *neutral*. In fact, we can quickly see that all points between the surface and B along the environmental curve are unstable. We say this is an *unstable layer*. Similarly, the points above B on the curve as shown form a *stable layer*.

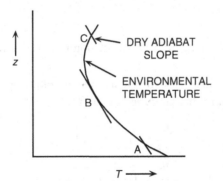

Figure 6.7 A sounding curve of the atmosphere: temperature T versus height z. Also indicated are straight line segments representing dry adiabats passing through the points A (unstable), B (neutral) and C (stable).

Obviously, the key to these analyses is the local slope of the environmental profile compared to the slope of a dry adiabat. Clearly

$$\left| -\frac{dT}{dz} \right| = \text{lapse rate} > \Gamma_d \rightarrow \text{unstable} \tag{6.42}$$

$$\left| -\frac{dT}{dz} \right| = \text{lapse rate} = \Gamma_d \rightarrow \text{neutral} \tag{6.43}$$

$$\left| -\frac{dT}{dz} \right| = \text{lapse rate} < \Gamma_d \rightarrow \text{stable}. \tag{6.44}$$

The vertical derivative of the potential temperature (see Section 3.3.3), $d\theta/dz$, provides an even simpler rule. The derivative of θ with respect to z can be calculated from its definition, $\theta = T (p/p_0)^{-\kappa}$ (with $\kappa = R_d/c_p$):

$$\frac{d\theta}{dz} = \frac{\partial\theta}{\partial T}\frac{dT}{dz} + \frac{\partial\theta}{\partial p}\frac{dp}{dz}$$

$$= \frac{\theta}{T}\frac{dT}{dz} - T\left(\frac{p_0}{p}\right)^{\kappa}\frac{\kappa}{p}\frac{dp}{dz}. \tag{6.45}$$

The hydrostatic equation can be used to simplify the second term:

$$\frac{d\theta}{dz} = \frac{\theta}{T}\frac{dT}{dz} + \frac{\theta}{p}\frac{R}{c_p}\rho g$$

$$= \frac{\theta}{T}\left(\frac{dT}{dz} + \frac{g}{c_p}\right)$$

$$= \frac{\theta}{T}\left(\frac{dT}{dz} + \Gamma_d\right). \tag{6.46}$$

This gives

$$\frac{d\theta}{dz} = \frac{\theta}{T}(\Gamma_d - \Gamma_e) \quad \text{[derivative of potential temperature]}. \tag{6.47}$$

In the last equation we introduced the local environmental lapse rate: $\Gamma_e = -dT/dz$. This last expression immediately tells us that

$$\frac{d\theta}{dz} > 0 \rightarrow \text{stable} \tag{6.48}$$

$$\frac{d\theta}{dz} = 0 \rightarrow \text{neutral} \tag{6.49}$$

$$\frac{d\theta}{dz} < 0 \rightarrow \text{unstable}. \tag{6.50}$$

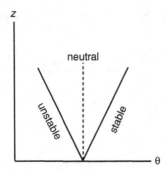

Figure 6.8 Schematic of three soundings on a θ–z diagram. The one with $d\theta/dz >$ 0 is stable; the neutral sounding has $d\theta/dz = 0$; the unstable one has $d\theta/dz < 0$.

In the θ versus z diagram, the dry adiabats are vertical lines (Figure 6.8). If θ is *increasing* with altitude, the layer is *stable*. If θ is decreasing with altitude, the layer is unstable. We will talk more about stability in Chapter 7, when we work with thermodynamic diagrams.

6.6 Vertical oscillations

Consider a level z_0 in a stable layer of the atmosphere such as point C in Figure 6.7. Since the layer is stable, a parcel displaced upwards will experience a restoring force tending to push it back to the point z_0. Similarly a downwards displacement below z_0 will result in an upwards restoring force. The acceleration a of a parcel slightly displaced along a dry adiabat at this point is given by Archimedes' Principle:

$$a = -(T_e(z) - T_a(z))\frac{g}{T_0} \tag{6.51}$$

where $T_e(z)$ is the environmental temperature profile or *sounding*; $T_a(z)$ is the local adiabat passing through the curve $T_e(z)$ at height z_0 (this is point C in Figure 6.7); T_0 is the value of the environmental temperature at the point of intersection, z_0, $T_0 \equiv T_e(z_0)$. Both environmental and adiabatic curves cross at this point. Recall that $dT_e/dz = -\Gamma_e$ and $dT_a/dz = -\Gamma_d$. Then the environmental and adiabatic temperatures near the point z_0 can be written as:

$$T_e(z) \approx T_0 - \Gamma_e(z - z_0) \tag{6.52}$$
$$T_a(z) = T_0 - \Gamma_d(z - z_0) \tag{6.53}$$

where we have used the approximate sign ($\approx$) to indicate that we are using only the tangent to the environmental curve; of course, the adiabatic lapse rate curve is a straight line, so the approximation is not necessary in the second equation. For notational convenience let $x = z - z_0$. Then we can equate the acceleration to the second time derivative of x:

$$\frac{d^2 x}{dt^2} = -(\Gamma_d - \Gamma_e)\frac{g}{T_0}x$$

$$= -\omega^2 x \qquad (6.54)$$

where

$$\omega^2 = (\Gamma_d - \Gamma_e)\frac{g}{T_0}. \qquad (6.55)$$

This is the familiar *harmonic oscillator* equation, whose solution is

$$x(t) = A\sin\omega t + B\cos\omega t \qquad (6.56)$$

where A and B are constants depending on the initial conditions,[4] and (in units rad s^{-1})

$$\boxed{\omega = \sqrt{g\frac{(\Gamma_d - \Gamma_e)}{T_0}}} \qquad (6.57)$$

This frequency is called the *Brunt–Väisälä* frequency. The frequency in cycles per second (Hz), $f = \omega/2\pi$, is

$$f = \frac{1}{2\pi}\sqrt{\frac{(\Gamma_d - \Gamma_e)}{T_0}g}. \qquad (6.58)$$

Physics refresher: oscillator notation The angular frequency of a linear oscillator is denoted by ω. Its units are radians per second. The corresponding frequency f is given by $\omega/2\pi$ in units of cycles per second or hertz. The period of the oscillation is $P = 1/f = 2\pi/\omega$.

When the atmosphere is stable ($\Gamma_d > \Gamma_e$) a parcel will oscillate; if the atmosphere is unstable, then $\omega^2 < 0$ and the trigonometric functions become a mixture of exponentials at least one of which is growing in time. To see this go back to the differential equation for the oscillator (6.54) and instead of $-\omega^2$ insert $\lambda^2 > 0$. Then the solutions become $Ae^{\lambda t} + Be^{-\lambda t}$. (You can insert this in the differential equation to satisfy yourself.) This means that instead of oscillating, the parcel "runs away." Another important observation is that for small amplitude oscillations (when the linear formula is valid) the frequency is independent of the amplitude of the displacement.

---

[4] We usually are not interested in the initial conditions of the oscillation, but rather the (angular) frequency, ω.

An alternative expression for ω in terms of the potential temperature is often useful. Referring to the last subsection we find:

$$\boxed{\omega^2 = \frac{g}{\theta} \frac{d\theta}{dz}}$$ [square of the Brunt–Väisälä frequency]. (6.59)

This particular representation shows explicitly that if $d\theta/dz$ is positive (stable atmospheric layer), then the quantity ω is a real number and the oscillations will occur. If the slope is negative, then the frequency becomes an imaginary number, which means that the parcel's displacement will grow exponentially in time, either up or down depending on the initial perturbation. Of course in the real atmosphere the parcel does not accelerate all the way to infinity, but instead its motion is limited by the failure of our linear analysis which assumed small deviations. Unstable layers lead to overturning and mixing of the parcels within the layer.

Example 6.10 Suppose the environmental lapse rate is $0\,\mathrm{K\,km^{-1}}$ (isothermal atmosphere) and $T_0 = 300\,\mathrm{K}$, then what is the oscillation frequency?
Answer:

$$f = \frac{1}{2\pi}\sqrt{\frac{\Gamma_{\mathrm{d}}}{T_0}g} = 0.0029\,\mathrm{s^{-1}} \tag{6.60}$$

which corresponds to a period of 5.7 min. □

It is typical of atmospheric vertical oscillations that the characteristic time is of the order of a few minutes.

6.7 Where is the LCL?

In this section we derive an approximate formula that shows that the LCL (lifting condensation level, see Section 5.9) is determined as a function of the temperature and mixing ratio, T_0 and w_0, of a parcel at the surface. Consider what happens to the saturation vapor pressure for this parcel. As the parcel rises its temperature falls from its surface value,

$$T_{\mathrm{a}}(z) = T_0 - \Gamma_{\mathrm{d}}z \tag{6.61}$$

where $T_{\mathrm{a}}(z)$ is used to denote the temperature of the parcel as it rises adiabatically to altitude z above the surface. The saturation vapor pressure only depends on the temperature inside the parcel. We can use the integrated form of the Clausius–Clapeyron equation (5.15) to evaluate the saturation vapor pressure $e_{\mathrm{s}}(T_{\mathrm{a}}(z))$ in the parcel as a function of altitude (along the dry adiabat):

$$e_{\mathrm{s}}(T_{\mathrm{a}}(z)) = e_{\mathrm{s}}(T_0)\exp\left(-\frac{L}{R_{\mathrm{w}}}\left(\frac{1}{T_{\mathrm{a}}(z)} - \frac{1}{T_0}\right)\right) \tag{6.62}$$

where $R_w = 461.5\,\mathrm{J\,kg^{-1}\,K^{-1}}$ is the gas constant for water vapor, and L is the enthalpy of vaporization (latent heat) for water ($L = 2.500 \times 10^6 \,\mathrm{J\,kg^{-1}}$). We can simplify the last equation by noting that

$$\frac{1}{T_a(z)} - \frac{1}{T_0} \approx \frac{\Gamma_d}{T_0^2} z \tag{6.63}$$

which leads to:

$$e_s(z) = e_s(T_0)e^{-z/H_w}, \quad \text{where } H_w = \frac{R_w T_0^2}{L\Gamma_d}. \tag{6.64}$$

As the parcel is lifted adiabatically (by some external mechanism) its mixing ratio, $w = w_0$, will remain fixed since it is a conservative quantity. The external air pressure can be written as an exponential function, $p_0 e^{-z/H}$, to a good approximation, where H is the atmospheric scale height. As a parcel rises adiabatically its internal pressure will adjust to that of the external pressure at each altitude z. Since $w_0 = \epsilon e(z)/p(z)$ we can obtain a formula for the vertical dependence of the actual vapor pressure $e(z)$ in the parcel

$$e(z) = \frac{w_0\, p(z)}{\epsilon} = 1.608\, w_0\, p_0\, e^{-z/H} \tag{6.65}$$

where the atmospheric scale height, H, can be taken to be $R_d T_0/g$ with T_0 the temperature near the surface. The value of H_w ($\approx 1.7\,\mathrm{km}$) is much smaller than typical values of H. An example is shown in Figure 6.9 where H was taken to be 8.3 km. At the surface $e(z) \ll e_s(z)$, but the gap narrows as the parcel is lifted adiabatically. The value of $e(z)$ will catch up with $e_s(z)$ as z increases. The saturation mixing ratio changes as the parcel rises, as shown in Figure 6.10.

Consider the T–z diagram shown in Figure 6.11 to see how the temperature of the parcel decreases linearly with altitude as it rises. At the same time the temperature versus height for $w_s = $ constant also decreases but more slowly. The intersection of the two curves is the LCL. This latest view is the one usually shown in thermodynamic diagrams which are the subject of the next chapter.

By equating the forms for $e_s(T(z))$ to that of $e(z)$ we find a formula for z_{LCL}

$$z_{LCL} = \frac{\ln\,(e_s(T_0)\epsilon/w_0 p_0)}{1/H_w - 1/H}. \tag{6.66}$$

We can simplify this further by noting that the argument of the logarithm reduces to $w_s/w_0 = 1/r$ where r is the relative humidity at the reference level (surface or sea level):

$$z_{LCL} = \ln(1/r)/(1/H_w - 1/H). \tag{6.67}$$

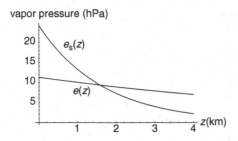

Figure 6.9 An example of how the saturation vapor pressure shrinks faster than the vapor pressure in a parcel rising from $z = 0$. The initial mixing ratio is 0.007 kg/kg and the initial saturation mixing ratio is 0.015 kg/kg. The initial temperature is 20 °C. The scale height of the atmosphere is taken to be 8.3 km and the scale height for the saturation vapor pressure is 1.66 km. Where the curves cross is the lifting condensation level (LCL).

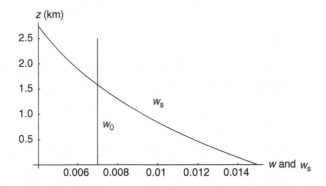

Figure 6.10 Illustration of how the mixing ratio ($w(z)$) and the saturation mixing ratio ($w_s(z)$) change as the parcel is lifted from $z = 0$. Note that $w(z) = w_0$ is a constant of the motion, but $w_s(z)$ decreases. Where the curves cross is the lifting condensation level (LCL). The values of the parameters are the same as in Figure 6.9.

This formula allows us to find an approximate value for the LCL for a given value of w_0 and T_0.

Example 6.11: profile of water vapor in the atmosphere Consider the profile of $e_s(z)$ in (6.64). Let us imagine that the water vapor in the atmosphere is saturated. We can use the results above to compute the ratio of $e_s(T(z))$ to the saturation value at the surface, $e_s(T(0))$. Let the relative humidity be 100% all the way up. The profile is shown in Figure 6.12 for the parameter values of the previous figures. Note that the scale height for the (saturated) water vapor is ≈ 1.66 km, which is a typical height of the atmospheric boundary layer. This is essentially the explanation of why water vapor is confined to the lowest few kilometers of the atmosphere. □

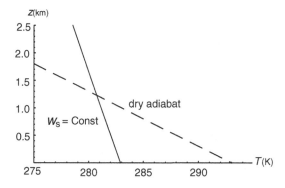

Figure 6.11 The vertical ascent of a parcel from $z = 0$ as a function of temperature. The curve labeled $(T - T_0)/\Gamma_d$ is the temperature of the parcel during dry adiabatic ascent; T_0 is the temperature at the surface. The curve labeled $w_s(T, p(z)) = $ constant is the saturation mixing ratio. Where the curves cross is the lifting condensation level (LCL).

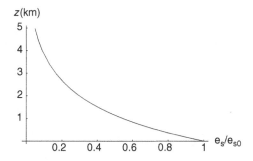

Figure 6.12 The ratio $e_s(T(z))/e_s(T(0))$ for water vapor embedded in an atmosphere with temperature profile $T(0) - \Gamma_d z$. The scale height in this example is $H_w = R_w T_0^2 / L\Gamma_d \approx 1.66$ km. The values of the parameters are the same as in Figure 6.11.

6.8 Slope of a moist adiabat

As was discussed in the previous section, an air parcel lifted adiabatically experiences a decrease in temperature at the dry adiabatic rate until the vapor in the parcel becomes saturated, defining the LCL. As the parcel continues to rise, water vapor will be converted to droplets with an accompanying warming of the parcel. This assumes the presence of condensation nuclei as discussed in Chapter 5. This is virtually always the case (but the number density of condensation nuclei differs significantly from place to place especially from above land surfaces to above ocean surfaces). This additional heating from condensation causes the temperature to decrease at a lower rate (as a function of altitude) compared to the dry adiabatic process. The slope of the moist adiabat can be found using arguments similar to those in the previous section.

The change in enthalpy for a parcel of mass $\mathcal{M}$ and volume V undergoing a small vertical displacement is given by

$$dH \approx \mathcal{M} c_p \, dT = dQ + V \, dp \tag{6.68}$$

where we use the approximate sign because we are neglecting some small terms. For example, we do not include the contributions from water vapor in the parcel $\mathcal{M}_v c_p^v$ as well as that of the liquid water droplets. The composite parcel rises with no heating from the outside such that $dQ = 0$, but if we treat the dry air as the only constituent of our system, there will be heating as water vapor is converted to droplets. We can write $dQ = -\mathcal{M} L \, dw_s$. Note that this is a positive number for rising air since $dw_s < 0$ (w_s decreases for the rising parcel). This last is because the saturation mixing ratio decreases with altitude as temperatures are lowered along the lift. Taking into account the hydrostatic equation (6.3) and dividing both sides of the equation (6.68) by $\mathcal{M}$, we rewrite (6.68) as

$$c_p \, dT = -L \, dw_s - g \, dz \tag{6.69}$$

or, after dividing both sides by $c_p \, dz$,

$$\frac{dT}{dz} = -\frac{L}{c_p} \frac{dw_s}{dz} - \frac{g}{c_p}. \tag{6.70}$$

Identifying $\Gamma_m = -dT/dz$ and $\Gamma_d = g/c_p$ leads to

$$\Gamma_m = \Gamma_d + \frac{L}{c_p} \frac{dw_s}{dz}. \tag{6.71}$$

The second term on the right-hand side has two parts since $w_s = w_s(T, p) = \epsilon e_s(T)/p$. Expanding the derivative:

$$\frac{dw_s}{dz} = \frac{\partial w_s}{\partial T} \frac{dT}{dz} + \frac{\partial w_s}{\partial p} \frac{dp}{dz}. \tag{6.72}$$

Next, insert the hydrostatic expression for dp/dz, using the dry air density. Then evaluate the partial derivative of $w_s(T, p)$:

$$\frac{dw_s}{dz} = \frac{\partial w_s}{\partial T}(-\Gamma_m) + \frac{\epsilon e_s}{p^2} \frac{pg}{R_d T}. \tag{6.73}$$

Now evaluate $\partial w_s/\partial T$ using the Clausius–Clapeyron equation (which introduces R_w, the gas constant for water vapor):

$$\frac{dw_s}{dz} = \frac{\epsilon}{p} \frac{L e_s}{R_w T^2}(-\Gamma_m) + \epsilon \frac{e_s}{p} \frac{g}{R_d T}$$

$$= \frac{w_s L}{R_w T^2}(-\Gamma_m) + w_s \frac{g}{R_d T}. \tag{6.74}$$

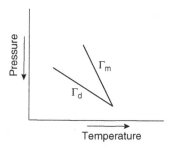

Figure 6.13 Relative slope of a dry adiabat (lapse rate Γ_d) and that of a moist adiabat (lapse rate Γ_m) started from the same point on a T–p diagram.

Then inserting this into (6.71) and rearranging, we have

$$\Gamma_m = \Gamma_d \left(\frac{1 + L w_s / R_d T}{1 + L^2 w_s / c_p R_w T^2} \right). \tag{6.75}$$

This rather cumbersome formula cannot be simplified further. The numerator in large parentheses is of the order of 1.3 and the denominator is of the order of 2.8 when typical numbers are inserted. This means that the moist adiabatic lapse rate is less than the dry adiabatic lapse rate (Figure 6.13).[5] This is hardly a surprise since the parcel is heated as it rises at saturation because of the condensation.

6.9 Lifting moist air

As we have just learned, an adiabatically rising parcel containing some moisture will eventually reach its condensation level. Above this point, with further lifting, water droplets form and grow at the expense of vapor in the parcel. As vapor is condensed, the parcel of air is heated, causing temperature changes along the vertical path, and thereby altering (actually increasing) the buoyancy of the parcel. If the water droplets are lost due to precipitation, it is referred to as a *pseudo-adiabatic* process. The prefix *pseudo* indicates that this process is irreversible (because of the precipitation) and cannot strictly be considered as an adiabatic process. Another possibility is that an ascending air parcel retains the water droplets after they are formed. Such a process is called a *moist* adiabatic process. Temperature differences between the inside of the parcel and that outside are approximately preserved because the molecular thermal conduction (and even the eddy transport) is too slow. Consistent with the other approximations we have adopted, the changes in the parcel's thermodynamic coordinates in a lifting process can be taken to be

---

[5] More detailed derivations retaining the contributions due to liquid and vapor can be found in the book by Bohren and Albrecht (1998) and that by Irebarne and Godson (1981). These complications add greatly to the tedium of the derivation, but the approximate result is in sufficient agreement to warrant omitting them here.

reversible and adiabatic. The actual difference between the pseudo-adiabatic and the moist adiabatic processes is negligible for a wide range of problems in atmospheric science. In this book we will not distinguish between pseudo-adiabatic and moist adiabatic processes. The path of a saturated air parcel will be called a *moist adiabat*.

It has already been shown that the path of a dry air parcel can be conveniently described in terms of the potential temperature, which is constant during a dry adiabatic process and, consequently, serves as a perfect *tracer* of adiabatic motion for a dry air parcel. We call it a tracer since we could label the parcel with its (fixed) potential temperature and follow it around. In this section we will introduce a new variable that is conserved during both dry and moist adiabatic processes, and can be used to trace a moist air parcel.

We start with the definition of entropy as applied to an ideal gas. Suppose the parcel is saturated. In an infinitesimal lift a tiny (positive) mass of water (per kilogram of air) equal to $-dw_s$ is condensed. The (positive) change in entropy for a moist air parcel with mass $\mathcal{M}$ is then $-\mathcal{M}L\,dw_s/T$. On the other hand, the change in entropy for a dry parcel of ideal gas is $\mathcal{M}c_p\,d\theta/\theta$ (see the discussion surrounding equation (4.34)). We can equate these two expressions for the infinitesimal entropy change, and after cancelling the common mass factor we have:

$$-\frac{L\,dw_s}{T} = \frac{c_p\,d\theta}{\theta}. \tag{6.76}$$

To a good approximation[6]

$$\frac{dw_s}{T} \approx d\left(\frac{w_s}{T}\right). \tag{6.78}$$

Then we have:

$$c_p\frac{d\theta}{\theta} = -L\,d\left(\frac{w_s}{T}\right). \tag{6.79}$$

Upon integrating each side:

$$-\frac{Lw_s(T,p)}{T} = c_p\ln\theta(T,p) + \text{constant} \tag{6.80}$$

[6] To show this consider the quantity

$$d\,(w_s/T) = \frac{w_s}{T}\left(\frac{dw_s}{w_s} - \frac{dT}{T}\right). \tag{6.77}$$

We can use the Clausius–Clapeyron equation to show that the first term in parentheses is much larger than the second term (recall that for every 10 K of increase in temperature there is a doubling of vapor pressure; then $dw_s/w_s \sim 1$, and $dT/T \sim 10/300$. Compare: $1 \gg 10/300$). We may then substitute $d(w_s/T)$ for dw_s/T to a good approximation.

where $\theta(T,p) = T(p_0/p)^\kappa$. This last relationship (6.80) forms an implicit functional relationship that defines a curve in the T–P plane. The relationship can also be written

$$\theta(T,p) = \theta_e \, e^{-Lw_s(T,p)/c_p T} = T\left(\frac{p_0}{p}\right)^\kappa \tag{6.81}$$

The coefficient in front of the exponential, θ_e, is called the *equivalent potential temperature*. The equivalent potential temperature is conserved along the path of a moist parcel.

For a dry adiabat $\theta_{dry}(T) = $ constant, but for the *moist* adiabat, $d\theta/dT > 0$. If we solve for θ_e from (6.81), we obtain

$$\theta_e = \theta(T,p)e^{\frac{Lw_s(T,p)}{c_p T}}. \tag{6.82}$$

To see the physical significance of θ_e let us lift the parcel until all its water is condensed out (this means $p \to 0$ or $z \to \infty$). In this limit $w_s \to 0$ in (6.82) and the equivalent potential temperature θ_e becomes equal to the potential temperature θ. In other words, to find an equivalent potential temperature, one should lift the air parcel until all moisture is condensed and precipitates out, then compress the dry parcel adiabatically downwards until it reaches 1000 mb. The temperature the parcel attains at the 1000 hPa level is the equivalent potential temperature θ_e. The whole process is supposed to occur without exchanging heat with the environment. Note that θ_e is a unique label that can be attached to any air parcel, given its values of T, w and p at a particular level.

If the parcel is initially saturated and has temperature T_0 at level p_0, the equivalent potential temperature θ_e can be calculated by substituting its temperature T_0, its potential temperature $\theta(T_0, p_0)$, and the saturation mixing ratio $w_s(T_0)$ into (6.82). If the parcel is initially unsaturated, then the temperature, potential temperature, and saturation mixing ratio are to be calculated at the lifting condensation level (LCL). Since the mixing ratio w is equal to the saturation mixing ratio w_s at the LCL, the formula for equivalent potential temperature for an unsaturated parcel becomes

$$\theta_e = \theta(T_{LCL}, p_{LCL}) \, e^{L w/c_p T_{LCL}}. \tag{6.83}$$

We emphasize again that the equivalent potential temperature is conserved during both dry and moist adiabatic processes, while potential temperature is conserved only during dry adiabatic processes. This is the reason for using θ_e: it can serve as a good tracer for a moving air mass. Imagine, for example, a moist flow that passes over a mountain. If air in the flow is initially unsaturated, it could be lifted by convection on the upslope side of the mountain to the LCL, where the condensation process starts. During the lifting moisture is removed by raining out on the upslope side. Then as the air descends back to the surface on the lee side of the mountain, it will be much warmer and drier than on the upslope side. This is the origin of the Chinook wind (more on this in Chapter 7). So, the temperature, potential temperature, and mixing ratio vary during both ascent and descent of air parcels. At the same time, the equivalent potential temperature is the same at the starting and ending points; it is conserved during this complicated process.

There is yet another good tracer of moist air: the so-called wet-bulb potential temperature. The *wet-bulb potential temperature*, θ_w, is the temperature an air parcel would have if cooled from its initial state adiabatically to saturation, and then brought to 1000 hPa by a moist adiabatic process. This algorithm of finding the wet-bulb potential temperature depends on whether or not the parcel is initially saturated. If the parcel is initially saturated, it should be carried along a moist adiabat to the 1000 hPa pressure level. If the parcel is initially unsaturated, it should be lifted first to the LCL and then taken moist adiabatically to the 1000 hPa level. When descending, an air parcel may need additional water vapor to maintain saturation. The wet-bulb potential temperature, like the equivalent potential temperature, is conserved during both dry and moist adiabatic processes. So, in the case of the Chinook wind it is the same on the upslope and lee sides of the mountain.

The last useful characteristic of moist air that we introduce in this section is the *saturation equivalent potential temperature* θ_s. Consider an unsaturated parcel. The saturation equivalent potential temperature is the equivalent potential temperature the parcel would have if it started out completely saturated. The saturation equivalent potential temperature θ_s can be defined as:

$$\theta_s = \theta e^{L w_s(T,p)/c_p T}. \tag{6.84}$$

It is important to understand the difference between (6.82) and (6.84). θ and w_s in (6.82) are the potential temperature and saturation mixing ratio of saturated air at temperature T, whereas the same variables in (6.84) are calculated at the temperature T of unsaturated air. The saturation equivalent potential temperature is not conserved during an unsaturated process. For saturated air, θ_e is equal to θ_s. The reason we introduce θ_s is that it is a useful characteristic of air flow when analyzing air stability (we will discuss it briefly in Chapter 7).

6.10 Moist static energy

We can find a very simple form for the enthalpy of moist air,

$$H = \mathcal{M}_d c_p T + L\mathcal{M}_w + \mathcal{M}_w c_{pw} T \tag{6.85}$$

where the second term represents the contribution due to the enthalpy of vaporization, and the third is the enthalpy of the vapor. If the water were to condense, the second term would contribute to raising the temperature. Neglecting the third term (which is very small compared to the others) the specific enthalpy can be written

$$h = c_p T + L w_s \tag{6.86}$$

where we have neglected the mass of water vapor compared to that of the dry air.

If we consider a parcel of air being lifted, there is actually another term due to the gravitational potential energy per unit mass gz, where g is the gravitational acceleration and z is the elevation above some reference level. Kinetic energy could also be added but we neglect it here. The sum of enthalpy and gravitational potential energy is conserved along a vertical path. This sum is called the *moist static energy*. The term static is used because we neglected kinetic energy. As the parcel is lifted, the moist static energy (call it h_{mse}) is conserved. Note that below the LCL this means that

$$\frac{dh_{mse}}{dz} = 0, \qquad \rightarrow \qquad \frac{dT}{dz} = -\frac{g}{c_p}, \tag{6.87}$$

which is the dry adiabatic lapse rate. Above the LCL we can obtain further information about the moist lapse rate. For example, the formula for the slope for the moist adiabat (6.75) can be derived from it.[7]

6.11 Profiles of well-mixed layers

Very often a layer of finite thickness is caused to be well mixed by turbulent processes (stirring). For example, in the first kilometer or two above the ground the air is turbulent due to mechanically driven eddies that are induced by the larger scale air flows interacting with the surface features. If the atmospheric profile is stable, buildings, trees and other protuberances above a flat boundary will cause irregularities of the air flow. Moreover, if the air is unstable, convective irregularities will add to the mechanical turbulence. This kind of turbulent, well-mixed layer may

---

[7] See Bohren and Albrecht (1998).

persist up to a few kilometers where it gently changes to the more orderly larger scale flow. The layers above this boundary layer are called the *free atmosphere*.

In a mixed layer as a whole we do not have a strict thermal equilibrium. That is to say the layer will not reach a uniform temperature as a function of height. The mechanical stirring overrides the tendency for the layer to come to a uniform temperature due to thermal conduction (due to molecular or eddie transport processes). The reason for this is that as parcels rise their temperatures are lowered because of adiabatic expansion. Observations show that for such well-mixed layers, especially near the ground and on gusty days, the temperature profile approaches the dry adiabatic lapse rate. An example is the layer between 850 hPa and 1000 hPa shown in Figure 7.13.

6.11.1 Well-mixed temperature profile

A heuristic proof of the adiabatic lapse rate in a well-mixed layer can be constructed by assuming that the atmosphere is subdivided into horizontal layers, each labeled by an index, i. Now suppose a piece of one of the lower layers is carried to a higher layer and in turn the same amount of mass from the upper layer is carried below by the mechanical stirring mechanism. As the parcel from below is lifted adiabatically and then brought into contact with the layer above, it will be in thermal contact with other parcels in that horizontal layer. The lifted parcel and the others in the upper level will reach a temperature (isobaric mixing) between the original environmental temperature of the layer and that of the adiabatically lifted parcel. The adjustment to equilibrium in this upper horizontal layer represents an increase in entropy of that layer which can be treated as a thermodynamic system. The collection of all the layers can be thought of as a collection of thermodynamic systems which we allow to interact in this peculiar way. Each time a parcel is lifted or lowered and brought into contact with a layer at a different pressure level, the entropy of that system increases and furthermore the entropy of the entire collection increases. As the mixing proceeds in this way, each step preserving the mass at an individual level and preserving the total enthalpy of the system, the system will come to a profile in which further mixing will no longer increase the entropy of the collection. This final state must be the one in which the entropy is homogeneous throughout. This is the state with constant potential temperature (recall $S = \mathcal{M}c_p \ln \theta$) and this is the adiabatic profile.

Mathematical derivation We can make our argument more compelling by using an analytical approach. First, take the entropy of the whole system of layers to be

$$S = \sum_i \mathcal{M}_i c_p \ln \theta_i. \tag{6.88}$$

We wish to preserve the total enthalpy of the composite system:

$$H = \sum_i \mathcal{M}_i c_p T_i = \sum_i \mathcal{M}_i c_p (\theta_i - \Gamma_d z_i). \tag{6.89}$$

We would like to find the extremum of S subject to the constraint that H be held constant. A convenient way to do it is through the use of a *Lagrange multiplier*, λ (almost all calculus books nowadays discuss this technique). We proceed by writing

$$W(\theta_1, \dots, \theta_n) = \sum_i \mathcal{M}_i c_p \ln \theta_i - \lambda \sum_i \mathcal{M}_i c_p (\theta_i - \Gamma_d z_i) \tag{6.90}$$

and set the partial derivatives to zero:

$$\frac{\partial W}{\partial \theta_j} = 0, \quad \frac{\partial W}{\partial \lambda} = 0. \tag{6.91}$$

This procedure finds the set of θ_i ($i = 1, \dots, n$) that will make S extreme. We find

$$\theta_i = \frac{1}{\lambda}, \quad \lambda = e^{-(S/\mathcal{M}c_p)z}. \tag{6.92}$$

In other words θ_i does not depend on i or z; it is a constant.

Of course, it must be kept in mind that the mathematical proof does not ease the assumptions we made about adiabatic lifting and lowering and the assumptions about horizontal (constant pressure) exchanges of heat between the parcel being moved and its environment at the same level (pressure). On the other hand, the fact that such a simple argument reproduces the profile seen in nature so regularly suggests that our assumptions are reasonable.

6.11.2 *Water vapor in a well-mixed layer*

We have remarked in earlier chapters that the mixing ratio, w, is a conserved quantity under vertical motions below the LCL. We can go through the same argument as above to show that the mixing ratio of water (or that of any other inert chemical species) should become uniform in the layer. Basically, when we bring a parcel into a layer in which the background is different from the mixing ratio in the parcel, the two will mix in such a way that the new mixing ratio will lie between that of the parcel and that of the whole layer in proportion to the masses. This mixing in an individual layer will increase the entropy of that layer. Each exchange of parcels will cause an increase in the entropy of the entire composite system until further exchanges do not increase the entropy. This final configuration will occur when the entire layer or composite of layers is at a uniform mixing ratio, w.

A uniform value of w in a layer exhibits a characteristic shape of the dew point temperature in a thermodynamic diagram. It turns out the dew point curve will lie exactly parallel to one of the saturation lines on the chart.

Ocean mixing There are layers in the ocean in which the mixing theory given above for the atmosphere works. The most important analogous conserved quantity in the ocean is the salinity. This quantity, along with potential temperature, is uniform in the deepest layers of the ocean where enough time has elapsed since their isolation to leave these water masses well mixed.

Notes

Many of the subjects in this chapter are covered in dynamics books such as Holton (1992). The thermodynamic details are discussed in more detail by Bohren and Albrecht (1998) and Irebarne and Godson (1981).

Notation and abbreviations for Chapter 6

A	horizontal area of a slab (m^2)
g	acceleration due to gravity ($9.81 \, m \, s^{-2}$)
$\Gamma_d, \Gamma_m, \Gamma_e$	lapse rate, $-dT/dz$ of dry air ascending adiabatically, of moist adiabat, of the environment ($K \, m^{-1}$)
h	height above a reference level
H	scale height
H_w	a scale height for water vapor (m)
$\kappa = R/c_p$	(dimensionless)
$L = \Delta H_{vap}$	enthalpy of vaporization (latent heat) ($J \, kg^{-1}$)
ω, f	angular frequency (rad^{-1}), frequency (Hz)
$p, p(z), p_0$	pressure, as a function of z, at a reference level (hPa)
$\Psi(z), \Psi_1, \Psi_2$	geopotential height as a function of height, at two levels (meters, on charts often in decameters, dm)
ρ, ρ_0, ρ_e	density, at a reference level, of the environment ($kg \, m^{-3}$)
$T, T(z), T_0$	temperature, as a function of z, at a reference level (K)
$T_e(z), T_a(z)$	temperature of the environment, of an adiabat (K)
$\overline{T}$	vertical average temperature in a layer of air (K)
$\theta, \theta_e, \theta_s, \theta_w$	potential temperature (K), equivalent potential, saturation equivalent potential, wet-bulb potential
w, w_s	mixing ratio, saturation mixing ratio (kg water vapor per kg dry air)
$z, \Delta z$	vertical distance, increment of it (m)
Z_p	potential energy per unit mass due to gravity (m)

Problems

6.1 Suppose the temperature of the atmosphere has the dependence $T = T_0 e^{-z/z_0}$. Find an expression for the pressure $p(z)$.

6.2 A $1\,\text{m}^3$ parcel of moist air ($r = 75\%$, $T = 303\,\text{K}$, $p = 1000\,\text{hPa}$) is embedded in surrounding dry air. What is the vertical acceleration of this parcel?

6.3 Suppose a parcel has a vertical acceleration of $0.12\,\text{m s}^{-2}$ (see previous problem). If it starts at rest at the surface, what is its vertical velocity after 5 s, 10 s, 30 s? How long does it take to reach the top of the boundary layer ($\approx 2\,\text{km}$)?

6.4 At a certain level of the (dry) atmosphere z, the temperature is 303 K and the local lapse rate is $12\,\text{K km}^{-1}$. Is this layer stable? Suppose a $1\,\text{m}^3$ parcel is displaced upwards by 0.5 km adiabatically. What is its acceleration due to buoyancy? How will the answer change if the parcel is displaced isothermally?

6.5 Suppose the atmosphere has its temperature equal to 300 K and pressure 1000 hPa at $z = 0$. The temperature profile falls linearly with a lapse rate of $6\,\text{K km}^{-1}$ up to 10 km. Above 10 km the temperature is constant. What is the pressure as a function of z?

6.6 Use the results of Problem 6.1 to compute the potential temperature as a function of height z.

6.7 Find the dry lapse rate near the surface for Mars. The mean radius of Mars $R_{\text{Mars}} = 3.40 \times 10^6\,\text{m}$ is $0.530 \times R_{\text{Earth}}$; mass of Mars $= 0.107 M_{\text{Earth}}$; universal gravitational constant $G = 6.67 \times 10^{-11}\,\text{N m}^2\,\text{kg}^{-2}$; for CO_2, $c_p \approx 0.76\,\text{kJ kg}^{-1}\,\text{K}^{-1}$.

6.8 Suppose an atmospheric profile is given by $T(p) = a + b\ln p/p_0, 0 < p \leq p_0$. Find an expression for the geopotential height $Z(p)$ as a function of pressure, p.

6.9 What is the thickness of the 1000 to 900 hPa layer if the mean temperature is 300 K?

6.10 What is the acceleration of a dry air parcel whose temperature is 300 K embedded in an environment of 285 K?

6.11 Compute the Brunt–Väisälä frequency for dry air in a layer where $d\theta/dz = 1\,\text{K km}^{-1}$, $\theta = 300\,\text{K}$. Give the answer in Hz as well. Compute the period of the oscillations.

6.12 Consider the differential equation:

$$\frac{d^2 x}{dT^2} = -\omega^2 x.$$

Show that

$$x = A\cos\omega t + B\sin\omega t$$

is a solution for constant values of A and B.

6.13 Relating the last problem to buoyant oscillations of a dry air parcel, find the coefficients A and B for two situations, using $d\theta/dz = 1\,\text{K km}^{-1}$ and $\theta \approx 300\,\text{K}$: (a) $x(0) = 10\,\text{m}$, $v(0) = 0\,\text{m s}^{-1}$; (b) $x(0) = 0\,\text{m}$, $v(0) = 1\,\text{m s}^{-1}$.

7

Thermodynamic charts

Atmospheric scientists make use of a variety of charts in their analysis of weather conditions. In the previous chapter we learned to identify whether a parcel located at a particular altitude is stable under small perturbations. The stability is determined by the local slope of the environmental curve compared to that of an adiabat passing through the same point. The diagram used in those studies was the temperature versus the altitude. On such a diagram a plot of the observed environmental temperature versus altitude could be compared with plots of hypothetical adiabatic trajectories of parcels. We learned that comparing the local slopes could reveal the stability of air located at a point (altitude) on the environmental curve.

While the temperature and the altitude are convenient for determining the local stability, the temperature and the logarithm of the pressure are more appropriate coordinates for computing energetic quantities of interest without giving up the convenient stability rules. Since the atmosphere is very nearly in hydrostatic equilibrium, the altitude, temperature and log pressure are closely related through the hydrostatic equation, $dp/p = -(g/RT)dz$. Using $\ln p_0/p$ (p_0 is a standard pressure, usually 1000 hPa) as the vertical coordinate instead of altitude will allow us to relate the energetics of a parcel's movements in an unstable environment.

To see how T versus $\ln p$ is related to thermodynamic diagrams used earlier in this book, consider a closed loop integral enclosing an area in this plane:

$$\oint T \, d\ln p = \oint \frac{T}{p} \, dp = \frac{1}{R} \oint v \, dp \tag{7.1}$$

where $v = 1/\rho$ is the specific volume of a parcel. We can write

$$v \, dp = -p \, dv + d(pv) \tag{7.2}$$

so that

$$\oint v \, dp = -\oint p \, dv + 0 \tag{7.3}$$

163

since $\oint d(pv) = 0$. This means the area enclosed by the loop integral in the T versus $\ln p$ plane is proportional to the negative of the work done *by* the gas in the parcel in traversing the loop. This means the area is proportional to the work *on* the gas by the environment. This provides a connection to the thermodynamic diagrams we are familiar with. We will see the physical significance and applicability of this connection in the next section.

7.1 Areas and energy

Consider a parcel perched at an unstable equilibrium point (altitude) in the atmosphere. A slight upwards nudge will cause the parcel to accelerate upwards because of the increasing (from zero) buoyant force. As the parcel rises away from its previous but precarious equilibrium level it will gain kinetic energy and lose buoyant potential energy. We can obtain a formula for the kinetic energy of the parcel as a function of distance above the initial level. We consider only adiabatic motion, wherein the air parcel moves without heat exchange with the surrounding environmental air.

According to Archimedes' Principle, the upward force per unit mass (equal to acceleration) on the parcel is

$$\frac{F}{\mathcal{M}} = -\frac{(\rho_a(z) - \rho_e(z))}{\rho_a(z)} g \tag{7.4}$$

where $\mathcal{M}$ is the mass of the parcel, $\rho_a(z)$ is the density of the air in the parcel as it moves vertically along an adiabatic path, and $\rho_e(z)$ is the density of the environmental air just outside the parcel at level z. Note that for displacements for which ρ_a is less than ρ_e there will be an upwards (positive) buoyant force on the parcel. The work done per unit mass on the parcel by the buoyancy force in moving the parcel from z_0 to z is $\int_{z_0}^{z} (F/\mathcal{M}) \, dz$, which is also the change in the kinetic energy per unit mass of the parcel in this displacement. In other words, the positive buoyancy force causes the parcel to increase speed as it moves in the vertical direction. Using $\mathcal{K}$ to denote the kinetic energy per unit mass, we obtain

$$\mathcal{K}(z) - \mathcal{K}(z_0) = -\int_{z_0}^{z} (\rho_a(z) - \rho_e(z)) \frac{g}{\rho_a(z)} dz$$

$$= \int_{z_0}^{z} (T_{a'} - T_e) \frac{g}{T_e} dz. \tag{7.5}$$

Substituting the hydrostatic equation, $dp/dz = -\rho g$, and the ideal gas equation of state, $p = \rho R T$, yields:

$$\mathcal{K}(z) - \mathcal{K}(z_0) = -R \int_{p_0}^{p} (T_a - T_e) d(\ln p). \tag{7.6}$$

The result is that the kinetic energy of a parcel is proportional to the area in the closed loop defined by doubly intersecting environmental and adiabatic curves in a T–$\ln p$ diagram:

$$\mathcal{K} = -R \oint T \, d \ln p. \tag{7.7}$$

It is worth remembering that the above derivation is valid for both dry and moist adiabatic processes. As the parcel rises adiabatically, its kinetic energy goes up, if there is a positive area enclosed by the parcel's path and environmental curve.

The same fact is, of course, true for the v–p diagram. The change of kinetic energy per unit mass in v–p variables is

$$\mathcal{K}(z) - \mathcal{K}(z_0) = -\int_{p_0}^{p} (v_a - v_e) \, dp$$

$$= -\oint v \, dp$$

$$= \oint p \, dv, \tag{7.8}$$

where we have substituted $v = v_a - v_e$.

We have seen that the v–p and T–$\ln p$ diagrams have the area-energy property: the area enclosed by the environmental curve on the left and an adiabatic (parcel) curve on the right is proportional to the kinetic energy per unit mass assumed by a parcel being forced upwards by buoyancy in such an unstable atmospheric profile.

Figure 7.1 shows a diagram of an environmental sounding curve with a dry adiabat leaving the surface and rejoining the sounding at a higher level in the atmosphere. Theoretically, a parcel leaving the surface along the dry adiabat will have a kinetic energy on reaching the environmental curve proportional to the shaded area bounded by the two curves.

Of course, the idea of frictionless motion of the parcel is highly idealized. Exchange of momentum transmitted by small eddies between the parcel and its environment tend to slow the parcel and alter its motion from the ideal conditions of frictionless motion. This *entrainment* process also exchanges other properties such as chemical composition and enthalpy. Nevertheless, the idealized kinetic energy parameter has proven useful in diagnosing and predicting the consequences of unstable situations.

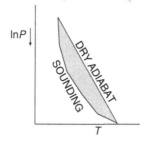

Figure 7.1 Schematic diagram of a sounding (left border of the shaded area) along with a dry adiabat rising from the surface (right border of the shaded area). The shaded area is proportional to the kinetic energy acquired by a parcel in rising from the surface to the intersection of the two curves.

We will return to energetic considerations after introducing the skew T diagram and passing through a series of chart exercises designed to build familiarity with the charts.

7.2 Skew T diagram

While in the previous section we saw that the T–ln p diagram has the very useful area-energy property, experience has shown that a related diagram is even more useful while preserving the area-energy property. While several such diagrams have been proposed over the years and discussions of them can be found on the internet, by far the most widely used is the skew T–log p chart which we will refer to simply as the skew T chart. Use of the diagram saves time, avoids tedious calculations, and provides an easy visual means of summarizing the vertical structure of the atmospheric thermal, stability, energetic and moisture characteristics.

Data from a *radiosonde* (an instrumented balloon launched every 12 hours at a network of thousands of locations over the globe) are plotted on the diagram to form the *sounding* or environmental curve. The main parameters derived from the radiosonde are the pressure, temperature, humidity, altitude and horizontal components of wind velocity.

The skew T diagram differs from the T–ln p diagram in that the abscissa is rotated about the origin ($T_0 \approx -50°C$, ln 1000) by about 45° downwards in a clockwise direction as illustrated in Figure 7.2.

The resulting coordinate plane (or diagram or chart) (Figure 7.3) is shown in the form used by practicing meteorologists and researchers. The abscissa X on this diagram is proportional to $(T + \beta \ln (p_0/p))$, where β is an adjustable coefficient set once and for all for convenience (for the usual skew T chart, the value is close to unity making the angle of rotation 45°). The ordinate Y is proportional to $\ln (p_0/p)$ which is very nearly proportional to the altitude, making interpretation

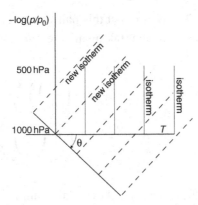

Figure 7.2 In the skew T diagram the T-axis is rotated about $45°$ clockwise. The original isotherms were vertical while the rotated ones are tilted as shown. The isobars are horizontal before and after the rotation (only the abscissa is rotated).

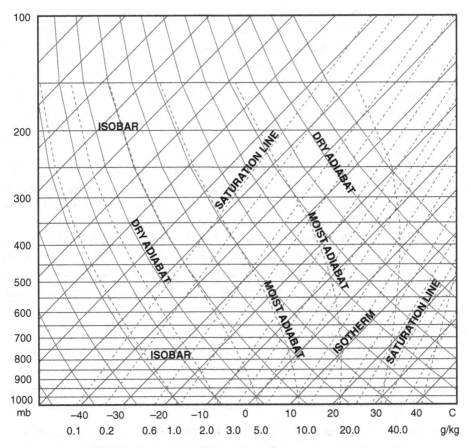

Figure 7.3 Illustration of a skew $T - \log p$ diagram. The isolines are labeled.

of this coordinate easier. To be sure that this pair of coordinates is viable from the point of view of energetics, consider taking an integral around a closed loop in the plane:

$$\oint X \, dY = \oint \left(T + \beta \ln \left(\frac{p_0}{p} \right) \right) \frac{dp}{p}$$

$$= \oint T \frac{dp}{p} - \beta \underbrace{\oint \ln \left(\frac{p}{p_0} \right) \frac{dp}{p}}_{=0}. \tag{7.9}$$

This tells us that a closed loop in a skew T diagram has the same value as that same loop in a T–ln p (unskewed) diagram. The larger the area enclosed the more energy will be involved in a related closed loop process. The closed loop process can then be related to the conversion of buoyant potential energy into kinetic energy of convection as we discussed in the last section.

In the skew T diagram the lines of constant pressure (isobars) are horizontal while the isotherms are no longer vertical but are tilted to the right. Examination of Figure 7.3 shows many lines besides the isobars and the isotherms. Let us take one curve at a time.

Dry adiabat We can obtain an equation for the dry adiabat by taking the logarithm of Poisson's equation,

$$\ln T = \ln \theta + \kappa \, \ln(p/p_0). \tag{7.10}$$

This relationship shows that the dry adiabats are not exactly straight lines on a skew T diagram. Hence, the dry adiabats are slightly curved (solid) lines which run from the lower right to the upper left of the diagram and are nearly perpendicular to the isotherms. The 45° angle between the skewed isotherms and isobars and the resulting 90° angle between isotherms and adiabats makes it easier for the observer to see the difference between the sounding and the adiabats. This property has led to the wide adoption of the skew T diagram. The pressure is in hPa (same as mb), the temperature is in degrees Celsius. The dry adiabats are labeled by the potential temperature associated with them.

Saturation mixing ratio These lines are drawn on the skew T diagram as dashed lines running toward the upper right. The units are g kg^{-1}, which indicate the amount of water in grams per kilogram of air at saturation at the particular temperature and pressure. The value for each saturation mixing ratio line is shown on the bottom of the diagram (on some charts on the internet it might be shown on the upper right).

Moist adiabat These are shown as dashed lines running toward the upper left. They can be computed from the information in the previous chapter, but this is unnecessary since the charts already provide the relationship. More about these below.

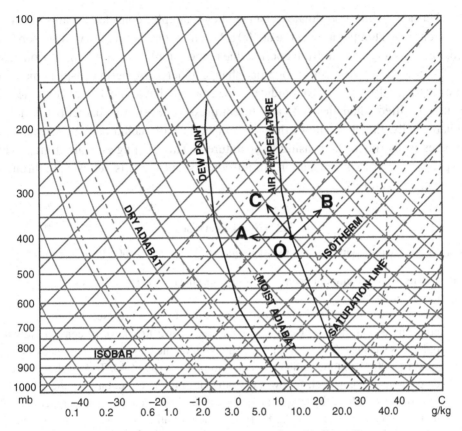

Figure 7.4 Example of a sounding on a skew $T - \log p$ diagram.

The thermodynamic state of a parcel of air can be represented as a point in the diagram. For example, a parcel having temperature $-10°C$ and pressure $600\,hPa$ is seen to have potential temperature $\approx 30°C$. To specify the parcel's properties completely we need to know its water vapor content, w in $g\,kg^{-1}$. Suppose our parcel which is located at $600\,hPa$ and $-10°C$ contains water vapor with a mixing ratio of $0.5\,g\,kg^{-1}$. These three quantities p, T, w are sufficient to define the thermodynamic state of the parcel. One could equally well specify p, θ, and RH since one triplet can be found from the other.

Next consider the sounding plotted in Figure 7.4. There are two sounding curves plotted on the chart: the temperature plot and the dew point plot. Both profiles are based on radiosonde measurements. The temperature line is always to the right of the dew point line. One can read the values of potential temperature, equivalent potential temperature, saturation mixing ratio, and actual mixing ratio of an air parcel situated at any particular pressure and temperature. A parcel of air can be moved hypothetically in different directions on the chart. As it moves its thermodynamic coordinates change. For example, consider a parcel on the environmental temperature curve at 400 hPa; a horizontal move (OA) is an isobaric change (this might result from cooling due to radiation). A move along an isotherm (OB) shows how the parcel's properties change under an isothermal displacement. A move along a dry adiabat (OC) indicates the changes that a parcel would experience if it were lifted adiabatically to a different pressure level. As a parcel is lifted dry adiabatically, the chart shows that the temperature of such a parcel decreases (it crosses isotherms of decreasing temperatures). This cooling is again a graphical expression of Poisson's equation. To introduce skew T charts, let us walk through some chart exercises.

7.3 Chart exercises[1]

Exercise 1 An air parcel has a temperature of 253 K at the 600 hPa pressure level.

(a) Find its potential temperature and saturation mixing ratio using both skew T–log p chart and formulas.

Answer: (Figure 7.5) On the skew T diagram that we use (see Figure 7.5), the temperature is in degrees Celsius; for that reason we first have to convert Kelvins to degrees Celsius ($T = -20°C$). Find the point on the diagram where the abscissa is equal to $-20°C$ and the ordinate is equal to 600 hPa (point A). To find the potential temperature of the parcel, follow a dry adiabat to the intersection with the 1000 hPa level (point B). Read the temperature at point B. This is the potential temperature of the parcel which is equal to 293 K. To find the saturation mixing ratio, from point A follow the line with constant saturation mixing ratio to the intersection with the abscissa (dashed line). Read the value of saturation mixing ratio (1.3 g kg^{-1}).

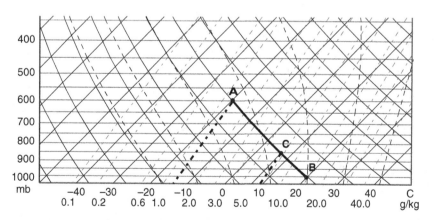

Figure 7.5 Diagram for Exercise 1.

[1] In these exercises we use reduced accuracy, $T_{STP} \cong 273$ K and $p_{STP} \cong 1000$ hPa as STP except when T_{STP} appears in the integrated Clausius–Clapeyron equation.

Using formulas, (i) find the potential temperature from Poisson's equation, (ii) find the saturation mixing ratio from the integrated form of the Clausius–Clapeyron equation (5.17) and formula (5.35).

(b) Move the parcel along the dry adiabat (dry adiabatically) to the 850 hPa level. What is the new temperature of the parcel? What is the new saturation mixing ratio? What is the parcel's potential temperature at 850 hPa?

Answer: From point A follow the dry adiabat to a pressure level of 850 hPa (point C). To find the temperature, follow the isotherm from point C to the intersection with the abscissa. Read the temperature ($T = 7°C$). To find the saturation mixing ratio, from point C follow the line of constant saturation mixing ratio to the intersection with the abscissa. Read the value of saturation mixing ratio ($7\,\mathrm{g\,kg^{-1}}$). The parcel's potential temperature at the 850 hPa level is the same as at 600 hPa, since the descent was conducted dry adiabatically with conservation of potential temperature.

Exercise 2 An air parcel has a temperature of 298 K at 1000 hPa level. Its mixing ratio is $14\,\mathrm{g\,kg^{-1}}$. Find the relative humidity using the chart and formulas.

Answer: Find the point on Figure 7.6 with abscissa 25°C and ordinate 1000 hPa corresponding to the initial conditions of the parcel (point A).

We know the mixing ratio, hence, to find the relative humidity we have to find the saturation mixing ratio (see Exercise 1). The saturation mixing ratio at point A is $20\,\mathrm{g\,kg^{-1}}$. Thus the relative humidity is

$$\mathrm{RH} = \frac{14\,\mathrm{g\,kg^{-1}}}{20\,\mathrm{g\,kg^{-1}}} \times 100\% = 70\%.$$

Using the formulas, the temperature of the parcel is known, therefore from the Clausius–Clapeyron equation we find the saturation pressure to be 32 hPa. Then, calculate the saturation mixing ratio from formula (5.35). Finally, find the relative humidity from (5.28).

Exercise 3 Using the same initial conditions as in Exercise 2, find the dew point of the parcel using both chart and formulas.

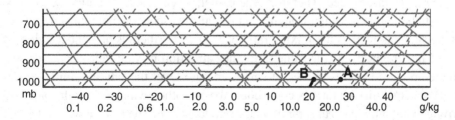

Figure 7.6 Diagram for Exercises 2 and 3.

Answer: By definition, the dew point is the temperature to which an air parcel must be cooled at the same pressure level in order for it to be saturated. In our case the parcel is unsaturated, its mixing ratio being less than the saturation mixing ratio. So, to find the dew point move to the left from the initial condition of the parcel (point A) along the isobar corresponding to the 1000 hPa level. Stop at the intersection of the isobar with the line of constant saturation mixing ratio corresponding to 14 g kg^{-1} (point B). This shift along the isobar corresponds to cooling the parcel at the same pressure level until its mixing ratio becomes equal to its saturation mixing ratio. Therefore, the temperature (the abscissa) at point B is the dew point. For this problem $T_D = 19\,°C$. It is important to understand that the dew point and mixing ratio of the parcel reflect equivalent information: if you know the dew point, you can find the mixing ratio at the same temperature and pressure, and vice versa. Now let us calculate the dew point using formulas instead of diagrams. To find the dew point, we have to equate the saturation mixing ratio to the actual mixing ratio of the parcel:

$$w_s(T_D) = w, \tag{7.11}$$

$$\frac{0.622 \times 2.497 \times 10^9 \, \text{hPa} \exp(-5417/T_D)}{1000 \, \text{hPa}} = 0.014. \tag{7.12}$$

This gives us the dew point temperature $T_D = 19\,°C$.

Exercise 4 An air parcel is lifted adiabatically from the 1000 hPa level where the parcel has a temperature of 20°C and dew point 6°C. Find the LCL (lifting condensation level). What are the temperature and potential temperature of the parcel at this level?

Answer: Find the parcel's initial location on Figure 7.7 corresponding to 20 °C abscissa and 1000 hPa ordinate (point A). The saturation mixing ratio at 20°C is 14.5 g kg^{-1}. Since we know the dew point, we can easily find the actual mixing ratio: the magnitude of the saturation mixing ratio at the dew point (6°C at the 1000 hPa level, point B) is the actual mixing ratio, which is equal to 5.7 g kg^{-1} in our case. Since the actual mixing ratio is less than the saturation mixing ratio, the air parcel is unsaturated. Therefore, the parcel, when lifted adiabatically, follows a dry adiabatic line passing through point A. During ascent the potential temperature is constant, as well as the mixing ratio of the parcel (there is no condensation and latent heat release). At the same time, the saturation mixing ratio decreases since the temperature decreases. So, at some pressure level the saturation mixing ratio and the actual mixing ratio become equal to each other, which means that adiabatic lifting eventually leads to saturation and condensation. The level at which the air in the parcel has cooled by adiabatic expansion sufficiently to become saturated is the

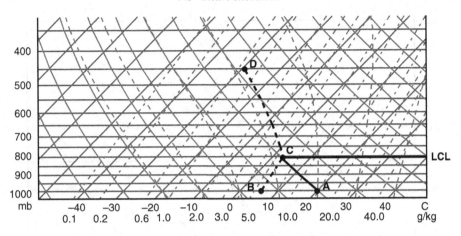

Figure 7.7 Diagram for Exercises 4 and 5.

LCL. The LCL can be found at the intersection of the dry adiabat starting from the initial parcel's temperature and pressure (AC on the graph) and constant saturation mixing ratio line starting from the dew point (BC). In our case the LCL (point C) is at 810 hPa. The temperature at this level is 3°C. The potential temperature is 293 K at the LCL. The potential temperature is conserved during the lifting; it is the same as that at the 1000 hPa level.

We can calculate the LCL without using charts, but this is more difficult. We have to find at what pressure level the dry adiabat that started at point A intersects with the line of constant saturation mixing ratio that started from point B. We can, for example, express temperature in terms of potential temperature and pressure using Poisson's equation and substitute it into the Clausius–Clapeyron equation. As a result, we have to solve a transcendental equation for pressure. Charts are quicker.

Exercise 5 Continue Exercise 4. If the parcel is lifted adiabatically to the 450 hPa level, what is its final temperature?
Answer: At 810 hPa the parcel has been saturated. So, its further lifting is along a moist adiabat. Follow the moist adiabat starting from point C to the 450 hPa level. The temperature at this point is −29°C.

Exercise 6 An air parcel at the 800 hPa level with temperature −10°C is saturated (mixing ratio 2.2 g kg^{-1}). Compute the equivalent potential temperature θ_e using both the skew T–log p chart and the formula for θ_e.
Answer: Lift the parcel from its initial location (point A on Figure 7.8) along the moist adiabat to infinity (approximately 200 hPa on the graph, moist and dry adiabats are nearly parallel to each other at this and higher levels). All vapor that the parcel initially contained has condensed. Then move the parcel back to the

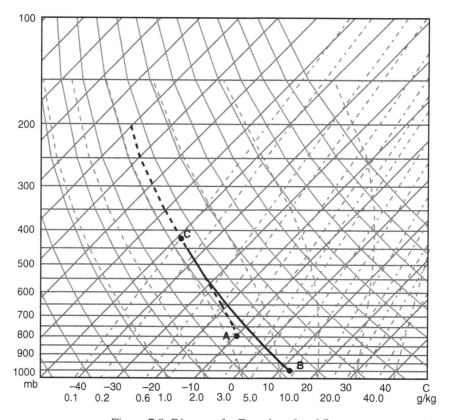

Figure 7.8 Diagram for Exercises 6 and 7.

1000 hPa level (point B) along the dry adiabat. The temperature at point B is the *equivalent potential temperature* (286 K).

From formula (6.82) we get

$$\theta_e = 263\,\text{K} \times \left(\frac{1000}{800}\right)^{0.286} \times \exp\left(\frac{2.5 \times 10^6\,\text{J kg}^{-1} \times 2.2 \times 10^{-3}}{1004\,\text{J kg}^{-1}\text{K}^{-1} \times 263\,\text{K}}\right) = 286\,\text{K}.$$

(7.13)

Exercise 7 Continue Exercise 6. Lift the parcel to the 425 hPa level. How much water is condensed during the ascent?

Answer: Bring the parcel from point A to the 425 hPa level along the moist adiabat (point C). The ascent is along the moist adiabat since the parcel was saturated at the initial temperature and pressure. At point C the saturation mixing ratio is equal to 0.10 g kg^{-1}. Hence, the amount of water condensed out during the ascent is 2.2 g kg^{-1} − 0.1 g kg^{-1} = 2.1 g kg^{-1}.

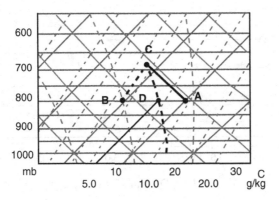

Figure 7.9 Diagram for Exercise 8.

Exercise 8 Air at 800 hPa has a temperature of 10°C and a dew point of 0 °C. Determine the wet-bulb temperature and the wet-bulb potential temperature.

Answer: The wet-bulb temperature, by definition, is the temperature an air parcel would have if cooled adiabatically to saturation at constant pressure by evaporation of water into it. To find the wet-bulb temperature on Figure 7.9 you have to perform the following steps. (1) Find the lifting condensation level (see Exercise 4). It is at 690 hPa (point C). (2) Draw a moist adiabat starting at C down to the intersection with the 800 hPa isobar (point D). Read the magnitude of the temperature at point D, which is the wet-bulb temperature (5°C). The wet-bulb potential temperature can be found by extrapolating the moist adiabat line starting at the LCL (point C) to the 1000 hPa level (288 K).

Exercise 9 The relative humidity of the air at the 950 hPa pressure level is 47%. Plot the relative positions of temperature, dew point and wet-bulb temperature on the chart.

Answer: The RH = 47% means that air is unsaturated. Therefore, the dew point (point B on Figure 7.10) is to the left of the temperature (point A) at the 950 isobar level. The wet-bulb temperature is always higher than the dew point. This happens because the dew point is the result of cooling to saturation at constant pressure with constant mixing ratio. Instead, the wet-bulb temperature characterizes cooling to saturation of the air parcel by evaporating water into the parcel, which raises the actual mixing ratio as the cooling proceeds. So, the wet-bulb temperature (point C) is between the dew point and the actual temperature.

Exercise 10: Chinook wind The Chinook wind is the warm dry wind that is the result of wind descending eastwards from the Rocky Mountains (Figure 7.11). Chinook winds can cause large temperature changes occurring only over a few

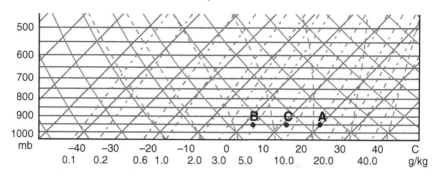

Figure 7.10 Diagram for Exercise 9.

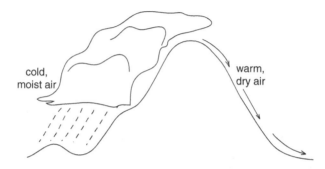

Figure 7.11 Exercise 10, illustration of the Chinook wind.

hours. "Chinook" is a Native American word that means "snow-eater," reflecting the warming effect, which could be accompanied by substantial melting. It is called Föhn in Europe. Let the air have a temperature of 10°C and 5 g kg^{-1} mixing ratio at pressure 950 hPa at the upslope (western) side of the mountain. When passing over the top of the mountain at the 600 hPa level, assume that 80% of the moisture is precipitated out. The air returns to the 950 hPa level on the eastern side of the mountain after being heated by the condensation. Compare the temperature, relative humidity, potential temperature and wet-bulb potential temperatures on both sides of the mountain at the 950 hPa level.

Answer: Find the initial location of the parcel on the chart (point A on Figure 7.12). At this point the saturation mixing ratio is equal to 8 g kg^{-1}. Since we know the mixing ratio of the parcel at point A, we can find the relative humidity RH = 62%. The potential temperature at point A is 287 K (see Exercise 1). The parcel is unsaturated at point A. Thus, when lifting, it follows a dry adiabat until it reaches the LCL (see Exercise 4). The LCL is at the 860 hPa level. The intersection of the dry adiabat starting at point A and the line of constant saturation mixing ratio of 5 g kg^{-1} is at point L. To find the wet-bulb potential temperature (see Exercise 8),

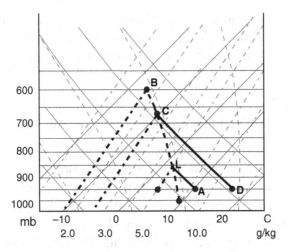

Figure 7.12 Diagram for Exercise 10.

draw a moist adiabat starting from L back to the 1000 hPa level. Read the abscissa, which yields 9°C. With further uplifting, the parcel follows the moist adiabat to the top of the mountain at the 600 hPa level (point B). At this point the parcel has saturation mixing ratio equal to 1.6 g kg^{-1}. Now we can find the amount of water condensed out during the ascent to the top of the mountain. It is equal to the difference between mixing ratios at points A and B, 5 g kg^{-1} − 1.6 g kg^{-1} = 3.4 g kg^{-1}. We know that 80% of this moisture, which is 2.7 g kg^{-1}, is precipitated out. Therefore, there is 3.4 g kg^{-1} − 2.7 g kg^{-1} = 0.7 g kg^{-1} of liquid water at the top of the mountain. When descending on the other side of the mountain, the air parcel follows the moist adiabat again since it is saturated. When descending, the parcel warms and expands, so water evaporates. Eventually, at some level, all the liquid water evaporates, the parcel is no longer saturated and its further descent is along a dry adiabat. How can we can find this "threshold" pressure level when the parcel reaches its new saturation mixing ratio corresponding to the evaporation of all liquid water? We know that at the top of the mountain there is 1.6 g of liquid water per kilogram of dry air. We also know that after the precipitation there is still 0.7 g of liquid water per kilogram of dry air. Hence, the new saturation mixing ratio, when all liquid water evaporates, is 1.6 g kg^{-1} + 0.7 g kg^{-1} = 2.3 g kg^{-1}. The parcel intersects the line with 2.3 g kg^{-1} saturation mixing ratio at the 660 hPa level (point C). This is the "threshold" pressure level because thereafter the parcel becomes unsaturated and follows its dry adiabat to the 950 hPa level (point D). The parcel's new temperature at point D is 17°C, which is 7 °C warmer than in the beginning. During dry adiabatic descent the mixing ratio is constant, so at point D the parcel has the same mixing ratio as at point C, which is 2.3 g kg^{-1}. The saturation mixing ratio at point D is 12.6 g kg^{-1}. Therefore, the relative humidity of the air on the other side of the mountain is 18% (compare with 62% at the beginning). The potential temperature at point D is 294 K (compare with 287 K at point A). The wet-bulb temperature is again 9 °C; it is conserved during the process.

7.4 Stability problem: example sounding

Figure 7.13 shows a typical sounding. In this example the air near the surface is unsaturated since the dew point is less than the air temperature. When lifted, the air follows a dry adiabat until it reaches the LCL as depicted in the graph. Further uplifting is along a moist adiabat (indicated by the dashed line). The first intersection of this moist adiabat with the sounding curve occurs at the *level of free convection* (LFC on the graph). If a parcel is lifted to a height lower than the LFC, it returns toward the surface because it experiences negative buoyancy since it is

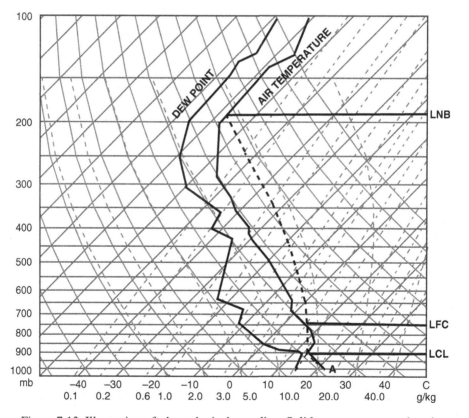

Figure 7.13 Illustration of a hypothetical sounding. Solid curves correspond to air temperature and dew point sounding, the dashed curve shows the actual path of a parcel lifting from the surface.

always cooler than the environment along its path. If, however, the parcel reaches the LFC, it becomes warmer than its surroundings. So, the LFC is the level where the parcel becomes positively buoyant. The positive buoyancy carries the parcel up to the *level of neutral buoyancy* (LNB), where the parcel's path intersects with the (measured) temperature sounding.

In Chapter 6 we discussed stability criteria for an unsaturated parcel. Let us apply these to different layers of the sounding. By layer we mean a thin slab of air along the sounding across which there is an approximately constant lapse rate. Consider the stability of layers depicted on Figure 7.13 (Figure 7.14 enlarges the part of the chart we are interested in).

Layer AB is stable since its lapse rate is less than the dry adiabatic lapse rate. Layer BC is a layer exhibiting a slight *inversion*. An inversion occurs in a layer when the temperature increases with height – such a layer is obviously stable. The layer CD is also stable, its lapse rate being less than the dry adiabatic lapse rate. The layer DE is neutral; it is parallel to the dry adiabat, so temperature decreases at the same rate as with a dry adiabatic process.

Now consider the case when the temperature decreases with height at a rate less than the dry adiabatic lapse rate but greater than the moist adiabatic lapse rate, $\Gamma_d > \Gamma > \Gamma_m$ (for example, layer AB). An air parcel in layer AB is negatively buoyant if lifted a short distance but could become positively buoyant if an imposed vertical motion is strong enough to bring this parcel to its level of free convection (LFC). For example, the air might be pushed up a mountainside or lifted by mechanically induced overturning (turbulence). Such a situation is called *conditional instability*. The layer is stable when air is unsaturated, but could become unstable with externally imposed vertical motion. We can test the layer for

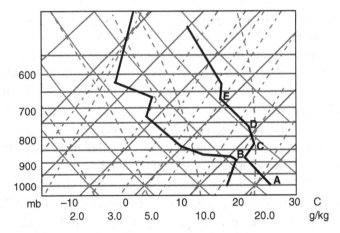

Figure 7.14 The enlarged area of interest of the sounding shown in Figure 7.13.

conditional instability by calculating the vertical gradient of saturation equivalent potential temperature $d\theta_s/dz$, rather than by calculating the lapse rate. In the case of conditional instability $d\theta_s/dz < 0$. Indeed, one can see from the graph (Figure 7.14) that $\theta_s(A) > \theta_s(B)$. If $\Gamma = \Gamma_m$ (A and B are on the same moist adiabat), then $\theta_s(A) = \theta_s(B)$, and $d\theta_s/dz = 0$.

To summarize, if the temperature in the particular layer decreases at a rate greater than the dry adiabatic lapse rate, this layer is unstable in any case for both saturated and unsaturated parcels. If the temperature decreases at a rate less than the moist adiabatic lapse rate, this layer is absolutely stable; the saturation equivalent potential temperature increases with height in this case. The formulas below list the stability criteria:

$$\Gamma > \Gamma_d \qquad \text{or} \qquad \frac{d\theta}{dz} < 0 \quad \text{absolutely unstable,} \tag{7.14}$$

$$\Gamma_d > \Gamma > \Gamma_m \qquad \text{or} \qquad \frac{d\theta_s}{dz} < 0 \quad \text{conditionally unstable,} \tag{7.15}$$

$$\Gamma < \Gamma_m \qquad \text{or} \qquad \frac{d\theta_s}{dz} > 0 \quad \text{absolutely stable.} \tag{7.16}$$

There is another type of instability called *potential instability*. Potential instability occurs when the layer is lifted as a whole, for example by convection associated with a moving front or with a flow passing over a mountain. When moving, the saturation conditions and, consequently, paths are different for the bottom and top of the layer, which can change the initial temperature gradient. Consider the inverted layer BC on the same sounding (Figure 7.15). The reason we chose an inversion layer is that the effect we want to demonstrate is more pronounced in this case. Imagine that an uplifting flow moves this layer as a whole 200 hPa higher. What happens? The bottom of the layer (point B) is almost saturated initially. Therefore, when lifted, it quickly reaches its LCL (labeled as LCL_B on the graph) and follows a moist adiabat thereafter (point B_1). The situation at the top of the layer (point C) is different. At the beginning, air at the top of the layer has a low relative humidity, its dew point is far to the left of its temperature. When uplifted, the air at point C reaches its LCL (labeled LCL_C on the graph) and then follows a moist adiabat (point C_1). Now consider the lapse rate of the B_1C_1 layer. It is larger than the dry adiabatic lapse rate. So, the absolutely stable layer BC becomes unstable when uplifted. This is the case of potential instability. The criterion for this instability is a negative gradient of the equivalent potential temperature in the layer, $d\theta_e/dz < 0$. Indeed, the equivalent potential temperature remains constant during the lifting everywhere: both above and below the LCL. When the air reaches the LCL, the equivalent potential temperature becomes equal to the saturation equivalent potential temperature, $\theta_e = \theta_s$. After that, from the criterion

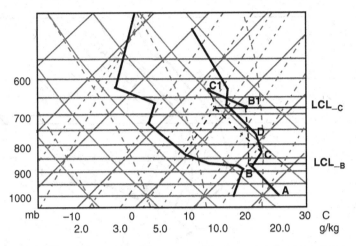

Figure 7.15 Illustration of potential instability. The stable layer BC becomes unstable after uplifting.

for conditional instability (7.15) we obtain $d\theta_e/dz < 0$ for potential instability. One can see from Figure 7.15 that $\theta_e(B) > \theta_e(C)$. This is a common occurrence in the southeastern USA as warm moist air is advected from the south, overriding drier air advected from the west. If lifting occurs, such a configuration can lead to severe weather conditions.

7.5 Convective available potential energy (CAPE)

In previous sections we analyzed the stability of the displacement of a small parcel in terms of temperature lapse rate. In this section we will continue to analyze stability, but in terms of energy. We have already shown that when there is a positive area in the closed loop between environmental and adiabatic curves on a T–ln p diagram or, in other words, if a parcel (after a nudge) is positively buoyant, the parcel's kinetic energy increases. Consider a parcel being initially unsaturated in a conditionally unstable atmosphere. We denote the parcel's initial location by A in the example of a temperature sounding shown in Figure 7.13. When lifted, the parcel first follows a dry adiabat until it reaches the LCL. With further lifting, it follows a moist adiabat. If the upward motion is strong enough to bring the parcel to its LFC, the parcel becomes positively buoyant. Figure 7.16 shows the same sounding as Figure 7.13. The positive area (shaded dark on Figure 7.16) between the parcel's path and the sounding bounded by the LFC and the LNB is called the *convective available potential energy* (CAPE). CAPE represents the maximum kinetic energy that a positively buoyant parcel can acquire by ascending without exchanging momentum (eddy friction), heat and moisture with its environment.

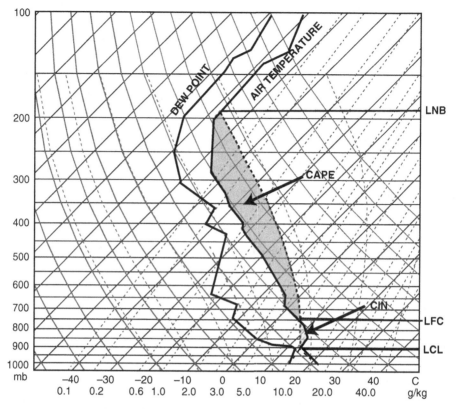

Figure 7.16 The same sounding as in Figure 7.13. Dark and light gray areas represent convective available potential energy (CAPE) and convective inhibition energy (CIN) correspondingly.

We can calculate the ideal change of kinetic energy per unit mass due to positive buoyancy by integrating (7.5) from LFC to LNB. The amount of kinetic energy released in this situation is

$$\text{CAPE} = \Delta \mathcal{K} = \int_{z_{\text{LFC}}}^{z_{\text{LNB}}} g \frac{T_a - T_e}{T_e} \, dz. \tag{7.17}$$

CAPE is a useful measure of thunderstorm severity, since it allows us to estimate the value of maximum possible vertical velocity. Indeed, if a parcel has zero vertical velocity at the LFC, then from (7.17)

$$w_{\text{max}} = \sqrt{2\text{CAPE}}. \tag{7.18}$$

In this consideration we have neglected the effect of water condensation, which reduces buoyancy slightly. Values of CAPE greater than $1000 \, \text{J} \, \text{kg}^{-1}$ imply the possibility of strong convection. Even if the final vertical velocity is less than the maximum value, the energy is still dissipated in turbulence within the cloud.

Let us return now to Figure 7.13. Before the parcel starting at point A reaches its LFC, it has to overcome a potential energy barrier between the LCL and the LFC, where a parcel is cooler than its environment and negative buoyancy tends to return the parcel toward the surface. This *negative* area between the parcel's path and environment bounded by the LCL and the LFC is called the *convective inhibition energy* (CIN). It is shown as the light gray area in Figure 7.16. CIN controls whether convection actually occurs. It is a measure of how much energy is required to overcome the negative buoyancy and allow convection. To find CIN we have to integrate (7.5) from the LCL to the LFC, namely

$$CIN = \int_{z_{LCL}}^{z_{LFC}} g \frac{T_a - T_e}{T_e} \, dz. \tag{7.19}$$

If the CIN is greater than $100 \, J \, kg^{-1}$, a significant source of lifting is needed to bring the parcel to its LFC in order to create favorable conditions for deep convection.

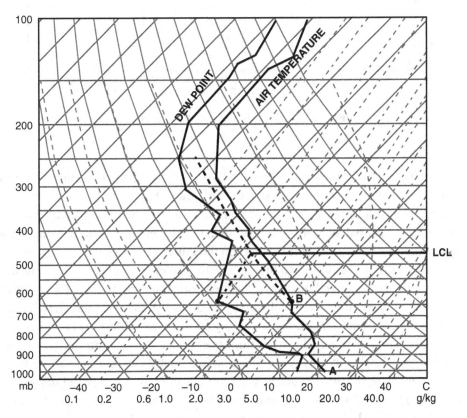

Figure 7.17 The same sounding as in Figure 7.13. For a parcel originating at point B, CAPE is zero.

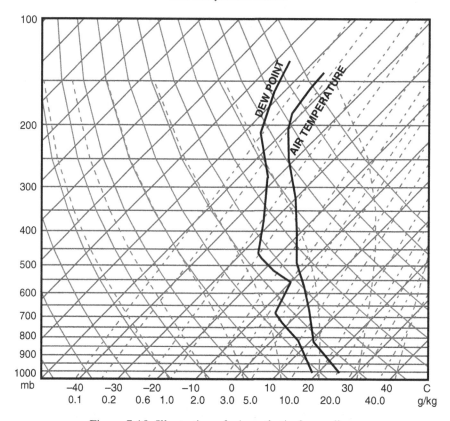

Figure 7.18 Illustration of a hypothetical sounding.

If we, for example, are interested in the CAPE of an air parcel starting at point B rather than at point A on the same sounding diagram (Figure 7.17), then the CAPE is zero. The path of the parcel starting at point B is shown by a dashed line on Figure 7.17. This parcel is always cooler than its local environment. It is important to note that the value of CAPE depends on the initial parcel location.

Consider the sounding shown in Figure 7.18. A parcel starting from the surface will experience negative buoyancy. The area corresponding to the CIN is shown in light gray in Figure 7.19, which simply enlarges the part of Figure 7.18 we are interested in. The area shaded in darker gray corresponds to the CAPE. To become positively buoyant, a parcel started from the surface (point A on the graph) has to overcome this "light gray" area. Imagine now that we expect the surface to be warmed in the next couple of hours. Then, instead of point A, the parcel starts from point A_1 (Figure 7.20). It does not experience negative buoyancy any longer; its LFC coincides with its LCL, and these are excellent conditions for severe thunderstorm activity. If, on the contrary, we expect the surface to be cooled (Figure 7.21, point

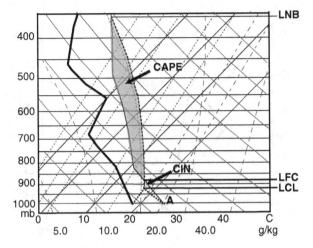

Figure 7.19 The same sounding as in Figure 7.18. CAPE and CIN for the parcel started from the surface (point A) are shown in dark and light gray correspondingly.

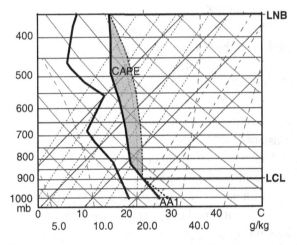

Figure 7.20 The same sounding as in Figure 7.18. Illustration of a hypothetical surface warming. For the parcel originating at point A_1, there is no CIN.

A_2), then the situation is reversed. CIN becomes larger, and CAPE is smaller than the previous situation. This means that the conditions for a thunderstorm are no longer favorable.

Fortunately for the forecaster, the values of many of the parameters discussed above (CAPE, CIN, etc.) are printed right on the skew T charts that are published at many sites on the internet. Hence, no tedious computations of areas are necessary by the user.

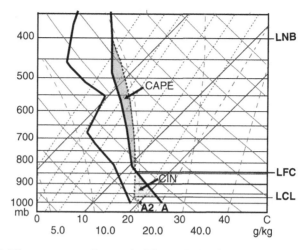

Figure 7.21 The same sounding as in Figure 7.18. Illustration of a hypothetical surface cooling. For the parcel originated from point A_2, CAPE decreases, but CIN increases in comparison with the parcel started from A.

Notation and abbreviations for Chapter 7

CAPE	convective available potential energy (J kg^{-1})
CIN	convective inhibition energy (J kg^{-1})
$F, F/\mathcal{M}$	force, per unit mass (N kg^{-1})
h, u	specific enthalpy, internal energy (J kg^{-1})
$\mathcal{K}$	kinetic energy (J)
L_{evap}	latent heat of evaporation (J kg^{-1})
R	gas constant $(\text{J K}^{-1}\,\text{kg}^{-1})$
ρ_a	parcel density for adiabatic change (kg m^{-3})
ρ_e	environmental density (kg m^{-3})
V_a, V_e	volumes of a parcel along an adiabat and of the environment (m^3)
w, w_s	mixing ratio, saturated (g water vapor per kg dry air)
X, Y	abscissa, ordinate

Problems

7.1 Refer to the sounding in Figure 7.22.
  (a) Estimate the mixing ratio at the surface.
  (b) Estimate the saturation mixing ratio at the surface. What is the relative humidity at the surface?
  (c) What is the dew point?
  (d) What is the pressure level at the LCL?
  (e) What is the wet-bulb temperature at the surface?
  (f) Is CIN > 0?

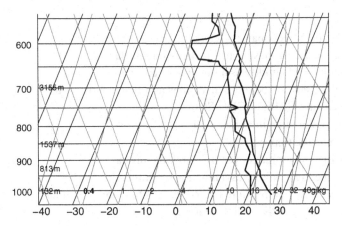

Figure 7.22 A sounding from Lake Charles, LA, at 00Z, June 1, 2007. Taken from the University of Wyoming website.

(g) Is there a large CAPE?

(h) What is the mixing ratio at 800 hPa?

(i) What is the saturation mixing ratio at 800 hPa?

7.2 Refer to the sounding in Figure 7.23.

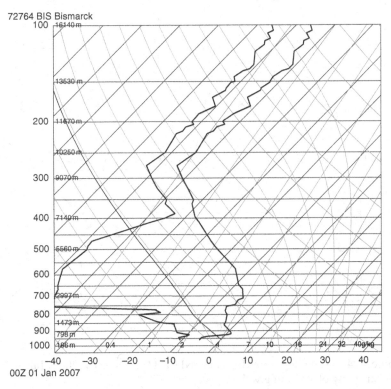

Figure 7.23 A sounding from Bismarck, ND, at 00Z, January 1, 2007. Taken from the University of Wyoming website.

(a) Where is the tropopause?

(b) Describe the air mass over Bismarck on this day.

(c) Describe the humidity as a function of altitude.

(d) Are there any temperature inversions as a function of altitude?

(e) Is there any CAPE or CIN? Stable?

7.3 Refer to the sounding in Figure 7.24.

(a) Where is the tropopause?

(b) Describe the air mass over Bismarck on this day.

(c) Describe the humidity as a function of altitude.

(d) Are there any temperature inversions as a function of altitude?

(e) Is there any CAPE or CIN? Stable?

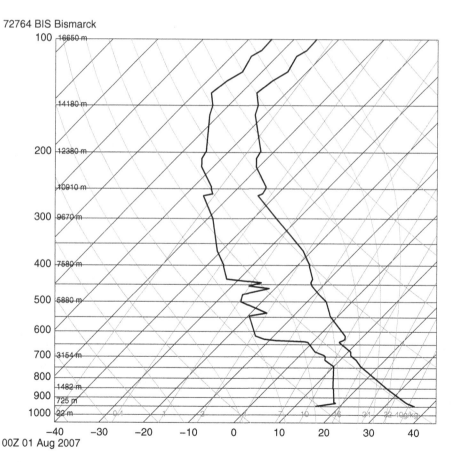

Figure 7.24 A sounding from Bismarck, ND, at 00Z, August 1, 2007. Taken from the University of Wyoming website.

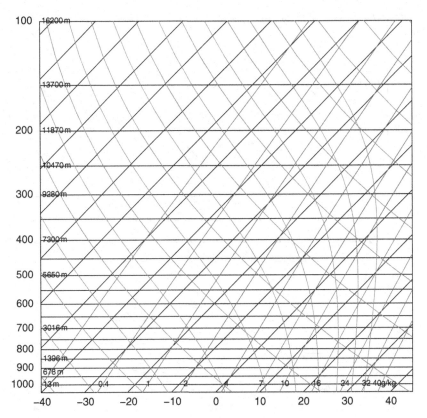

Figure 7.25 A blank skew T chart for Problem 7.4. Taken from the University of Wyoming website.

7.4 The air at 1000 hPa and 11°C has dew point -0.5 °C (see Figure 7.25).
   (a) Find the mixing ratio, relative humidity, and the potential temperature using both the skew T chart and formulas.
   (b) Find the lifting condensation level using the chart.
   (c) Find the equivalent potential temperature using the chart.
   (d) What are the mixing ratio and the potential temperature if the parcel rises to 900 hPa?
   (e) What is the equivalent potential temperature if the parcel rises to 600 hPa?

7.5 Consider a parcel of moist air that rises from the surface where $p = 1000$ hPa to 400 hPa. Assume all of the condensed water is precipitated out during the ascent. The parcel then descends (unsaturated) back to the surface. If the initial temperature is 20°C and its initial dew point is 0°C, find the following (use Figure 7.26).
   (a) How much water is condensed during the ascent?
   (b) The temperature of the parcel and its dew point temperature when it returns to the surface (1000 hPa).

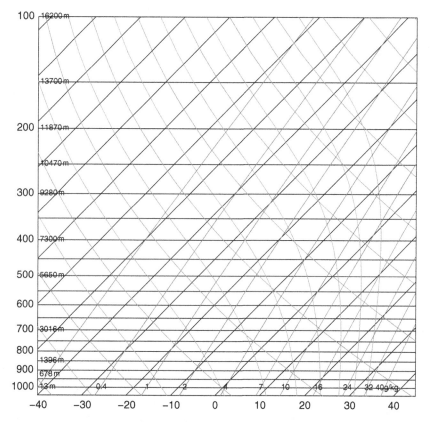

Figure 7.26 A blank skew T chart for Problem 7.5. Taken from the University of Wyoming website.

8

Thermochemistry

Applications of thermodynamics form the heart of *physical chemistry*. With the First Law of Thermodynamics we can find some elementary applications of thermodynamics to processes relevant to the atmosphere. As an example, consider the elementary reaction

$$CO_2(g) + H_2(g) \rightarrow CO(g) + H_2O(g) \qquad (8.1)$$

where the g in parentheses indicates that the chemical species is in the gaseous state (the solid phase is indicated by s, liquid by l and aqueous solution by aq).

At the molecular level a molecule of CO_2 strikes an H_2 molecule and the rearrangement collision occurs with a certain probability depending on velocities, spatial orientation of the colliders, etc. The rate at which a reaction proceeds in a system is the product of the likelihood of a collision between the important parties and the probability of rearrangement, given the collision. Sometimes a CO molecule bumps into an H_2O molecule and the reverse reaction occurs. That the reaction might go both ways is indicated by the equation

$$CO_2(g) + H_2(g) \rightleftharpoons CO(g) + H_2O(g). \qquad (8.2)$$

Once equilibrium is established (rate of reactions proceeding to the right equals the rate of those going to the left) in a suitable enclosure, we can consider the matter involved to be a thermodynamic system which can be treated by the methods of equilibrium thermodynamics. The number of moles of the species $\nu_{CO_2}, \nu_{H_2}, \nu_{CO}$ and ν_{H_2O} become thermodynamic coordinates or functions along with those we are already acquainted with, $\mathcal{M}, p, V, U, H, S, G,$ and T. In fact, we want to know how these coordinates (equilibrium concentrations) vary as a function of the temperature if the pressure is held constant. Fixed pressure is the usual condition for gas phase reactions in the atmosphere since they can be taken as occurring in a small parcel or volume element, whose pressure inside quickly adjusts to that outside (which in our case depends on altitude). In the real atmosphere there are many chemical species in various states of equilibrium. The task of the atmospheric chemist is often to sort out which reactions are important, what the sources of the various species are, what is the feasibility and energetics of chemical reactions and how fast the reactions proceed.

8.1 Standard enthalpy of formation

Consider a general chemical reaction

$$a A + b B \rightarrow c C + d D, \tag{8.3}$$

where A and B are called the *reactants*, C and D are the *products*, and a, b, c and d are integers (sometimes rational numbers) inserted to balance the equation.

Suppose the ingredients on the left-hand side of the equation (the *reactants*) are placed in a closed container that is impermeable to matter crossing its bounding surface. Furthermore, let the reaction (8.3) proceed from left to right at constant pressure. If no heat is allowed to enter or leave the system during the (irreversible) process, the final temperature will be different from that before the reaction began. If the temperature of the system goes up, we say the reaction is *exothermic*. If the temperature goes down, it is *endothermic*. Chemists have found a convenient way of characterizing the energetics of such reactions. Suppose the reaction goes from left to right to completion (no reactants remaining), then the heat required to restore the system to its original temperature at constant pressure is its change in enthalpy during the irreversible process, ΔH.

In order to find the heat of reaction for a particular chemical process it is necessary to start with the so-called *standard enthalpy of formation* of the individual compounds. These are based upon the enthalpy needed to form the compound from the state of the individual atomic species most commonly found in nature. For example, the convention for the element oxygen is to start with the gaseous form O_2, not O. Similarly the base state according to the convention for nitrogen is N_2 and for hydrogen it is H_2. For argon it is the atomic form Ar and for carbon it is C.

The standard enthalpy of chemical reaction, when reactants in their standard state are converted to products in their standard states, is equal to the difference between standard enthalpy of formation of products and reactants:

$$\Delta \overline{H}^\circ = [c\Delta \overline{H}^\circ(C) + d\Delta \overline{H}^\circ(D)] - [a\Delta \overline{H}^\circ(A) + b\Delta \overline{H}^\circ(B)]$$

$$= [\text{products}] - [\text{reactants}]. \tag{8.4}$$

The overbar indicates that 1 mol of the substance is considered, the superscript o refers to the standard state, which is at 1 atm and 25 °C by convention (see Table 8.1). If $\Delta \overline{H}^\circ$ is negative, heat is released and the reaction is exothermic. Exothermic

Table 8.1 *Standard enthalpies of formation for selected compounds* ($\Delta \overline{H}^{\circ}$ *in units of kJ mol^{-1}*)
The symbol in parentheses after the compound indicates whether its physical state is liquid, solid or gas. All values relate to 298 K.

$CO_2(g)$	-393.51	$CO(g)$	-110.53
$CH_4(g)$	-74.81	$H(g)$	$+217.97$
$H_2O(g)$	-241.82	$H_2O(l)$	-285.83
$O_2(g)$	0	$O(g)$	$+249.17$
$O_3(g)$	$+142.7$	$OH(g)$	$+38.96$
$HNO_3(g)$	-135.09	$NO_2(g)$	$+33.19$
$NO(g)$	$+90.25$		

reactions can proceed spontaneously in the atmosphere. If the opposite is true, $\Delta \overline{H}^{\circ}$ is positive, the reaction is endothermic, and an external source of energy is needed for the reaction to proceed.

The exothermic reactions can be significant for the thermal budget of the atmosphere. The classical example is the reaction leading to the formation of ozone. The heat released in this process dominates the form of the temperature profile in the stratosphere.

Example 8.1 The main mechanism of ozone formation in the stratosphere is the recombination of atomic oxygen:

$$O + O_2 + M \rightarrow O_3 + M, \qquad (8.5)$$

where M is a molecule in the background gas which is needed to carry off the excess momentum in a two-bodies-to-one molecular collision. Find how much heat is released by this reaction.

Answer: To find how much heat is liberated, we need to calculate the enthalpy of the reaction:

$$\Delta \overline{H}^{\circ} = [\Delta \overline{H}^{\circ}(O_3) + \Delta \overline{H}^{\circ}(M)] - [\Delta \overline{H}^{\circ}(O) + \Delta \overline{H}^{\circ}(O_2) + \Delta \overline{H}^{\circ}(M)]. \quad (8.6)$$

Since $\Delta \overline{H}^{\circ}(O_2) = 0$,

$$\Delta \overline{H}^{\circ} = \Delta \overline{H}^{\circ}(O_3) - \Delta \overline{H}^{\circ}(O). \qquad (8.7)$$

From Table 8.1, $\Delta \overline{H}^{\circ} = 142.7$ kJ mol^{-1} $- 249.17$ kJ mol$^{-1} = -106.4$ kJ mol^{-1}. The minus sign indicates that this is an exothermic reaction. Therefore, with the reaction of ozone formation (8.5) 106.4 kJ per mole is liberated. This liberated heat warms the stratospheric air and raises its temperature which reaches a maximum at about 50 km altitude. Note that the concentration of O in (8.5) is determined by the photodissociation of O_2, O_3 and other species.   □

Example 8.2 Suppose we wanted to know the change in enthalpy for the reaction:

$$CO_2 + H_2 \rightarrow CO + H_2O. \tag{8.8}$$

We form

$$\Delta\overline{H}^\circ = \Delta\overline{H}^\circ(CO) + \Delta\overline{H}^\circ(H_2O) - \Delta\overline{H}^\circ(CO_2) - \Delta\overline{H}^\circ(H_2)$$
$$= +(-110.53) + (-241.82) - (-393.51) - (0.0) \ (\text{kJ mol}^{-1})$$
$$= 41.16 \ \text{kJ mol}^{-1}.$$

$\Delta\overline{H}^\circ$ is positive, which means that this reaction is endothermic, and heat is absorbed during the process. □

8.2 Photochemistry

Further examples of endothermic reactions include the *photochemical reactions*. In this case the additional source of energy necessary for the endothermic reaction to proceed is solar radiation which can break the chemical bonds of atmospheric species. In this book we will consider only one photochemical process: *photodissociation*.[1]

Physics refresher Solar radiation consists of electromagnetic waves. Electromagnetic radiation has a dual wave-particle nature. This means that electromagnetic radiation exhibits both wave-like and particle-like properties. In its wave form electromagnetic radiation can be thought of as a group of superimposed waves sometimes referred to as an *ensemble* propagating in vacuum with the speed of light $c = 2.998 \times 10^8$ m s^{-1} independent of wavelength. Each wave in this ensemble can be treated as a simple sinusoidal function (see Figure 8.1) with a certain wavelength, frequency, and amplitude. The wavelength, λ, is the distance between two successive peaks of the wave. The units of λ are meters. The frequency of a wave, f, is the number of cycles that pass an observer in a second. The unit of frequency is the hertz (1 Hz is one oscillation per second). The product of wavelength and frequency for an individual wave is equal to the speed of light (speed is distance divided by time): $c = \lambda \times f$. From this equation one can see that waves with higher frequencies have shorter wavelengths, and waves with lower frequencies have longer wavelengths.

When radiation interacts with atoms or molecules, it can be absorbed or emitted only by certain discrete amounts of energy. In other words, electromagnetic radiation is *quantized*. The waves may be thought of as a beam of particles called photons carrying discrete amounts or packages of energy. The energy of a photon of

---

[1] Interested readers are referred to *Basic Physical Chemistry for the Atmospheric Sciences* by Peter V. Hobbs (2000) for more information on photochemical reactions.

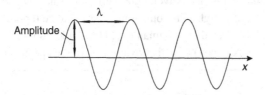

Figure 8.1 A sinusoidal wave of a given wavelength.

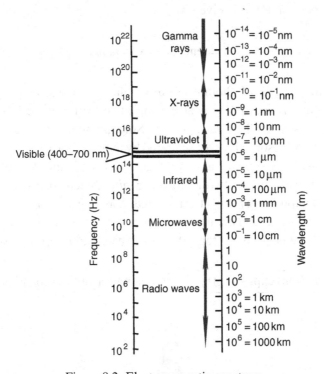

Figure 8.2 Electromagnetic spectrum.

frequency f is

$$E = hf \tag{8.9}$$

where h is *Planck's constant*, $h = 6.62 \times 10^{-34}$ J s. If we express f in terms of λ we obtain

$$E = \frac{hc}{\lambda}. \tag{8.10}$$

Hence, the shorter wavelength (higher frequency) photons are more energetic. The electromagnetic spectrum from radio waves to gamma rays is illustrated in Figure 8.2. The most energetic photons are gamma rays. As one moves vertically down the spectrum from gamma rays towards radio waves, the energy decreases and

Thermochemistry

so does the frequency, while the wavelength increases. A narrow band of the spectrum corresponds to the visible light. Photons in the visible range (approximately 400–700 nm) can be detected by a human eye. Ultraviolet radiation has shorter wavelength (higher frequency) than the visible part of the spectrum.

To describe the radiation penetrating the atmosphere it is useful to introduce the idea of an *energy flux*. The energy flux (energy passing per unit area perpendicular to the beam, per unit time) is given by

$$F = \text{energy flux} = n_0 h c f \qquad (8.11)$$

where n_0 is the number of photons per unit volume (*number density* as in a gas). The energy flux of solar photons at the top of the atmosphere is 1370 W m^{-2}. This parameter is called the *solar constant*. Solar photons propagating through the atmosphere can be absorbed and/or scattered by atmospheric constituents. Consider the *attenuation* of a photon flux at wavelength λ due to photon absorption assuming normal incidence for simplicity (the sun is at *zenith*, directly overhead (Figure 8.3)). We denote a photon energy flux at wavelength λ as F_λ; its dimension is energy per unit area, per unit time, and per unit wavelength. If at the top of the atmosphere the flux per unit wavelength is $F_\lambda(\text{top})$, the flux at height z, $F_\lambda(z)$, is described by

$$\boxed{F_\lambda(z) = F_\lambda(\text{top}) \exp(-\tau(z))} \quad \text{[attenuation of a vertical solar beam].} \quad (8.12)$$

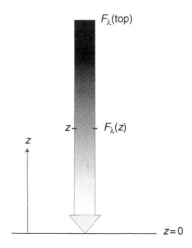

Figure 8.3 Schematic diagram of a solar beam coming from directly overhead with attenuation of the beam's intensity indicated by shading.

This equation follows from the radiative transfer theory. [2] The coefficient τ in the exponent is called the optical depth:

$$\tau = \sigma \int_z^\infty N(z)dz \quad \text{[optical depth].} \tag{8.13}$$

The optical depth τ is proportional to the vertically integrated column density $\int_z^\infty N(z)dz$ where N is a concentration of atmospheric species absorbing at λ. The integration from z to ∞ reflects the path the photons travel from the top of the atmosphere to height z. The coefficient of proportionality σ is called the absorption cross-section. This parameter describes the ability of a particular gaseous species to absorb photons; it is measured in m^2 (often in the literature as cm^2). Absorption cross-sections can be measured in the laboratory. When the optical depth gets close to unity, the flux is attenuated by a factor of roughly three (e $\approx$ 2.7). For example, for λ between 240 and 300 nm (ultraviolet range) τ reaches unity due to the absorption by ozone approximately at heights of 30–38 km. This means that the solar photons in this range are absorbed by ozone in the stratosphere and do not reach the troposphere. At shorter wavelengths, between 175 and 200 nm, radiation is absorbed by oxygen at heights of 40–80 km. At wavelengths greater than 310 nm, most photons penetrate into the troposphere and reach the surface. If the sun has zenith angle $\Theta \neq 0$, then $\cos \Theta$ has to be added in the formula for the flux attenuation (Figure 8.4):

$$F_\lambda(z) = F_\lambda(\text{top}) \exp(-\tau(z)/\cos\Theta) \quad \text{[attenuation at zenith angle } \Theta\text{].} \tag{8.14}$$

The larger the zenith angle, the stronger the attenuation at a given height z.

The photons in the ultraviolet and visible ranges are energetic enough to break molecules apart. This process is called photodissociation. Photodissociation plays a very important role in the troposphere and the stratosphere. For example, a key reaction in the troposphere is the photodissociation of ozone by ultraviolet radiation:

$$O_3 + hf \rightarrow O_2 + O \tag{8.15}$$

where the notation hf denotes a photon with frequency f. This notation emphasizes that the energy carried by the photon is the frequency times Planck's constant. The formation of tropospheric ozone is due to photodissociation of NO_2:

$$NO_2 + hf \rightarrow NO + O. \tag{8.16}$$

---

[2] A beam is attenuated in a distance interval dz by an amount proportional to the incoming beam's flux and to the amount of attenuating material in the interval. The result is $dF_\lambda = -AF_\lambda \, dz$ where A is proportional to the amount of attenuating material per unit volume. Integration of this equation leads to exponential decay along the path, known as Beer's Law.

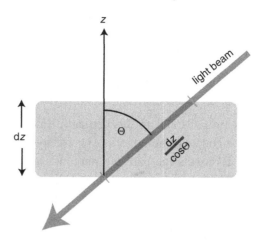

Figure 8.4 Schematic diagram of a solar beam coming from the upper right and passing through a slab of matter with thickness dz. The path length in the slab is dz/ cos Θ, where Θ is the solar zenith angle.

The atomic oxygen then leads to the formation of ozone by recombination with O_2:

$$O + O_2 + M \rightarrow O_3 + M. \tag{8.17}$$

The formation of the ozone layer is also caused by photodissociation. In the stratosphere, ultraviolet radiation with $\lambda \leq 240$ nm photodissociates molecular oxygen O_2 creating atomic oxygen:

$$O_2 + hf \rightarrow O + O. \tag{8.18}$$

Ozone is then formed by recombination of atomic and molecular oxygen (reaction (8.17)).

Example 8.3 Consider the photodissociation of an oxygen molecule that creates two ground state oxygen atoms:

$$O_2 + hf \rightarrow O + O. \tag{8.19}$$

What photon energy is required for this reaction to proceed? What part of the electromagnetic spectrum corresponds to this energy?
Answer: First let us examine the standard enthalpy of this reaction:

$$\Delta \overline{H}^\circ = 2 \times 249.17 - N_A\, hf = 498.34\,\text{kJ mol}^{-1} - N_A\, hf \tag{8.20}$$

where N_A is Avogadro's number. To find the minimum energy of a photon required to break one O_2 molecule we have to equate the standard enthalpy of this reaction to zero. Only photons with energy hf greater than this minimum energy are able to break an O_2 molecule apart: $hf \geq 498.34/N_A$ kJ. This inequality gives the value of the smallest frequency required: $f \geq 498.34 \times 10^3/(6.022 \times 10^{23} \times 6.62 \times 10^{-34}) = 12.49 \times 10^{14}$ Hz, or the largest wavelength. $\lambda \leq 0.24 \times 10^{-6}$ m = 240 nm. Therefore, radiation with $\lambda \leq 240$ nm, which corresponds to the ultraviolet part of the spectrum, is needed for the reaction (8.19) to proceed. □

We can find the energy of a photon necessary for a certain reaction to proceed without examination of the enthalpy, if we know the energy of dissociation of a chemical bond.

Example 8.4 During the daytime an important source of NO in the stratosphere is the dissociation of NO_2 molecules:

$$NO_2 + hf \rightarrow NO + O. \tag{8.21}$$

Find the maximum wavelength of electromagnetic radiation required for this reaction, if the energy of dissociation of an NO_2 molecule is 5.05×10^{-19} J. Energy of dissociation is often given in electronvolts:[3]

$$1 \text{ electronvolt (eV)} = 1.6 \times 10^{-19} \text{J} \tag{8.22}$$

or

$$5.05 \times 10^{-19} \text{J} = 3.16 \text{ eV} \tag{8.23}$$

(this reaction is also important in polluted urban air, since it is a source of tropospheric ozone).

Answer: Photons with energy $hf \geq 5.05 \times 10^{-19}$ J are needed to dissociate an NO_2 molecule. Then, $f \geq 5.05 \times 10^{-19}/(6.62 \times 10^{-34}) = 7.6 \times 10^{14}$ Hz. Finally, $\lambda \leq 2.998 \times 10^8/7.62 \times 10^{14} = 0.39 \times 10^{-6}$m = 390 nm, which is at the boundary between the visible and ultraviolet parts of the spectrum. □

8.3 Gibbs energy for chemical reactions

Earlier we showed how to find the enthalpy for phase changes and chemical reactions by manipulating values taken from standard tables. In this section we

---

[3] The unit electronvolt (eV) is the energy an electron has after being accelerated across a potential difference of 1 volt. This is the preferred unit in atomic and nuclear physics. The binding energy of an electron in the ground state of a hydrogen atom is 13.6 eV.

will work with the Gibbs energy, which is useful in determining the feasibility of a chemical reaction and the abundances of species in chemical equilibrium situations. Using enthalpy to determine whether a reaction will proceed is limited, since enthalpy depends on entropy S and pressure p as well as the concentrations of the various species present. If we are to examine whether a reaction will proceed, we will find it hard to hold the entropy constant, especially in nature. On the other hand, the Gibbs energy is useful when the temperature and pressure are held constant. This is often the case in the atmosphere when the reaction occurs between trace gases at a certain altitude (pressure) and the temperature is constant because the reagents are buffered thermally by the surrounding background gas molecules.

The *standard Gibbs energy* is introduced similarly to the standard enthalpy of the reaction. The standard Gibbs energy of a chemical compound, $\Delta \overline{G}^{\circ}$, is the change of the Gibbs energy when 1 mol of a compound is formed (the overbar is an indication of 1 mol being considered). Conventionally, the standard Gibbs energy of compounds in their most stable form is taken to be zero. The superscript o indicates the standard state, which is at 1 atm and 25 °C.

For the general chemical reaction

$$a \, A + b \, B \rightarrow c \, C + d \, D \tag{8.24}$$

the standard Gibbs energy is the difference between the Gibbs energies of products and reactants:

$$\Delta \overline{G}^{\circ} = [c \, \Delta \overline{G}^{\circ}(C) + d \, \Delta \overline{G}^{\circ}(D)] - [a \, \Delta \overline{G}^{\circ}(A) + b \, \Delta \overline{G}^{\circ}(B)]. \tag{8.25}$$

In Chapter 4 we learned that if temperature and pressure are held constant, then as the system tends spontaneously to its equilibrium, its Gibbs energy will decrease to a minimum. Applying this equilibrium criterion to chemical systems, we conclude that if $\Delta \overline{G}^{\circ}$ of the reaction is negative, the reactants in their standard state are are converted to the products in their standard state. If, on the other hand, $\Delta \overline{G}^{\circ}$ is positive, then an additional source of energy is needed for the reaction to proceed.

Example 8.5 Calculate the standard Gibbs energy of formation at 25 °C and 1 atm for the reaction:

$$HO_2 + NO \rightarrow NO_2 + OH. \tag{8.26}$$

Can it proceed spontaneously?

Table 8.2 *Standard Gibbs energy for selected compounds ($\Delta\overline{G}^\circ$ in units kJ mol^{-1}), all values relate to the standard conditions 298 K and 1 atm of pressure*

H_2O	-228.6	O_3	$+163.2$
OH	$+34.23$	HO_2	18.41
HNO_3	-74.79	NO_3	$+115.9$
NO_2	$+51.30$	NO	$+86.6$

Answer: The standard Gibbs energy of this reaction (Table 8.2)

$$\Delta\overline{G}^\circ = \Delta\overline{G}^\circ(NO_2) + \Delta\overline{G}^\circ(OH) - \Delta\overline{G}^\circ(HO_2) - \overline{G}^\circ(NO)$$
$$= (51.3 + 34.23 - 18.41 - 86.6)\,kJ\,mol^{-1} = -19.5\,kJ\,mol^{-1}.$$

Since $\Delta\overline{G}^\circ$ is negative, the reaction (8.26) can proceed spontaneously. Note that there is no information about how long the reaction will take to complete.   □

Example 8.6 Suppose we are looking for some effective mechanism of OH production in the atmosphere. We suggest that the recombination of H_2O and O_2 can work as a source for OH:

$$H_2O + O_2 \rightarrow HO_2 + OH. \tag{8.27}$$

Before we start the laboratory experiments to check our idea, we can calculate the Gibbs energy of this reaction:

$$\Delta\overline{G}^\circ = \Delta\overline{G}^\circ(HO_2) + \Delta\overline{G}^\circ(OH) - \Delta\overline{G}^\circ(H_2O). \tag{8.28}$$

After substituting the numbers from Table 8.2, we get $\Delta\overline{G}^\circ = 281.3$ kJ mol^{-1}. The positive value of standard Gibbs energy means that the suggested mechanism for OH formation is thermodynamically impossible in the atmosphere.   □

8.4 Elementary kinetics

We have seen how to estimate the energetics and feasibility of chemical reactions proceeding one way or the other using the methods of equilibrium thermodynamics. But equilibrium thermodynamics cannot tell us how rapidly a reaction will proceed.

This is the business of chemical kinetics which considers the details of the molecular collision and the intermediate complexes that can form during the event. For example, kinetics can provide a means of computing the characteristic time of the decay of the reactants in the atmosphere. One should keep in mind that negative Gibbs energy change for a reaction (thermodynamically favorable conditions) does not always mean that the reaction will proceed fast enough to be observed.

8.4.1 Reaction rate

A reaction rate can be defined intuitively as the rate at which the products of the reaction are formed, which is the same as the rate at which the reactants are consumed. As an example, consider a bimolecular reaction with molecules C and D as products and A and B as reactants:

$$A + B \rightarrow C + D. \tag{8.29}$$

The rate of this reaction (the rate of loss of A or B and the rate of increase of C and D) is

$$-\frac{d[A]}{dt} = -\frac{d[B]}{dt} = \frac{d[C]}{dt} = \frac{d[D]}{dt} = k[A][B], \tag{8.30}$$

where [X] denotes the concentration of species X expressed in molecules cm^{-3} and k is the *reaction rate coefficient*. The units of k depend on the order of the reaction: for the bimolecular reaction (8.30) k is in $cm^3 \, s^{-1}$. The reaction rate coefficient k is unique for a given reaction at each given temperature. The temperature dependence $k(T)$ is described by the *Arrhenius equation*:

$$\boxed{k(T) = A \, \exp\left(\frac{-E_{act}}{R*T}\right)} \quad \text{[rate coefficient with activation energy].} \tag{8.31}$$

E_{act} is called the *activation energy*. A large value of E_{act} usually implies a strong temperature dependence of the reaction rate coefficient. The constant A (not to be confused with the identity of the species A in (8.29)) before the exponential function is related to the frequency of molecular collisions and the probability for molecules to have an orientation in space favorable for a reaction. The dependence of A on temperature is usually weak compared to that of the exponential factor.

The idea of activation energy is shown schematically in Figure 8.5. The horizontal axis represents the reaction coordinate for the reactants. The reaction coordinate can be thought of as the distance between the molecules A and B in the reaction (8.29). The vertical axis is the potential energy of the reaction. $\Delta \overline{H}^{\circ}$ is the standard enthalpy of formation for this reaction. Note that $\Delta \overline{H}^{\circ}$ is negative, so the reaction is exothermic.

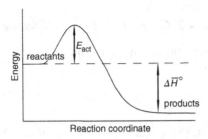

Figure 8.5 Schematic graph of energy change for an exothermic reaction. Reactants have to be energetic enough to overcome the barrier E_{act}.

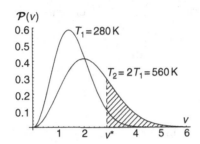

Figure 8.6 Velocity distribution of molecules at two different temperatures. The velocity v is expressed in units of $\sqrt{k_B T/m_0}$. The velocity v^* corresponds to the kinetic energy equal to E_{act}.

For the products C and D to be formed by this reaction, the reactants A and B must have enough kinetic energy to overcome the energy barrier E_{act}. It should be noted that many reactions in the real atmosphere do not proceed because the activation barrier is too large. For example, $C + O_2 \rightarrow CO_2$ does not take place in the atmosphere because of the large barrier.

Equation (8.31) implies that reactions proceed faster at higher temperatures.[4] This can be explained with the help of kinetic theory. It follows from (2.26) that the higher the temperature of the gas, the greater the fraction of molecules that have kinetic energies that exceed a certain given energy. Figure 8.6 shows the velocity distribution for two temperatures. At higher temperatures more molecules have velocities higher than the threshold velocity v^* corresponding to the kinetic energy $\frac{1}{2}m_0v^{*2}$ which is equal to E_{act}. This means that increasing the temperature of the gas increases the probability that molecules will overcome the barrier E_{act} and that the products will be formed at a higher rate.

---

[4] For some reactions the activation energy is actually negative (no barrier). The rate of these reactions decreases with increasing temperature.

8.4.2 Concept of chemical lifetime

Consider the case of a first-order reaction, when one element, A, decomposes into two elements, C and D:

$$A \rightarrow C + D. \tag{8.32}$$

The rate of decrease of the concentration of element A is proportional to its concentration:

$$\frac{d[A]}{dt} = -k[A] = -\frac{d[C]}{dt} = -\frac{d[D]}{dt}, \tag{8.33}$$

where k is the reaction rate coefficient. After rearranging the terms in (8.33) we have

$$\frac{d[A]}{[A]} = -k \, dt \tag{8.34}$$

and

$$\ln [A] = -k \, t + \text{constant}. \tag{8.35}$$

If at the initial time $t = t_0$ the concentration of A is equal to $[A]_0$, then

$$[A] = [A]_0 \, e^{-k t}. \tag{8.36}$$

This equation shows that the concentration of A decays exponentially with characteristic time $t_c = 1/k$. The time t_c required for the concentration of A to decrease by a factor of e from its initial value is called the chemical lifetime. The larger the reaction rate coefficient k, the shorter the lifetime t_c.

Example 8.7: half life The characteristic time for a unimolecular decay is t_c. What is the half life, i.e., what is the time after which half the concentration remains? We have

$$[A] = [A]_0 \, e^{-t/t_c}. \tag{8.37}$$

Then

$$\frac{1}{2} = e^{-t_{1/2}/t_c} \Rightarrow -\ln 2 = -t_{1/2}/t_c \Rightarrow t_{1/2} = \ln 2 \, t_c = 0.6931 \, t_c. \tag{8.38}$$

$$\square$$

Consider next the bimolecular reaction:

$$A + B \rightarrow C + D. \tag{8.39}$$

The rate of loss of A is given by

$$-\frac{d[A]}{dt} = -\frac{d[B]}{dt} = \frac{d[C]}{dt} = \frac{d[D]}{dt} = k[A][B], \qquad (8.40)$$

where k is the constant for this bimolecular reaction (8.39). An important case is when the concentration [B] is much larger than [A], then [B] can be considered a constant, say $[B_0]$ in the last equation. For example, gas A might be a trace gas such as atomic oxygen O, and gas B might be a background gas such as O_2 or N_2 (see e.g., (8.17)). This leads to

$$[A] \approx [A]_0\, e^{-k[B_0]t} \qquad (8.41)$$

and the lifetime of A is:

$$t_c = \frac{1}{k[B_0]}. \qquad (8.42)$$

For a photochemical reaction

$$A + hf \rightarrow C + D \qquad (8.43)$$

the decay of the concentration of molecule A is given by

$$\frac{d[A]}{dt} = -J[A] \qquad (8.44)$$

where J is the photodissociation coefficient expressed in s^{-1}. The photodissociation coefficient J in the interval $\Delta\lambda$ at height z is determined by the flux of photons with wavelength λ at height z, $F_\lambda(z)$, and the absorption cross-section [5] of molecules absorbing near λ, $\sigma(\lambda)$. Note that $F_\lambda(z)$ is the number of photons per unit area, per unit time, per unit wavelength (units of photons $m^{-3}\,s^{-1}$):

$$F_\lambda(z) = F_\lambda(\text{top})\exp(-\tau(z)/\cos\Theta)$$

(a discussion of $F_\lambda(z)$ can be found in Section 8.2). Integrating over a band of wavelengths $\Delta\lambda$ (we assume each photon striking a molecule dissociates it), the photodissociation coefficient for that wavelength band is

$$J(z) = \int_{\Delta\lambda} \sigma(\lambda)F_\lambda(z)d\lambda. \qquad (8.45)$$

One can see from (8.44) that the photochemical lifetime is the inverse J:

$$t_c = 1/J. \qquad (8.46)$$

---

[5] A typical value of $\sigma(\lambda)$ for the photoabsorption in the visible wavelength range by NO_2 is 5×10^{-5} nm^2 (Seinfeld and Pandis, 1998).

The concept of lifetime is useful in many atmospheric problems. Myriads of atmospheric constituents undergo chemical reactions and photochemical processes caused by solar radiation. In addition, chemical species are advected by transport processes in the atmosphere. Separation of processes with different time scales can simplify the problem significantly. In some cases this is the only way to analyze the variability in the very complicated world of atmospheric constituents. Suppose we know that the photochemical lifetime of a certain constituent is much smaller than the characteristic time of atmospheric transport at a given height. This is the case, for example, for stratospheric ozone: at altitudes higher than 30 km the chemical lifetime of ozone is several orders of magnitude smaller than the transport time scale. This allows us to neglect the effect of transport in the first order of approximation when analyzing ozone variability. Now consider methane in the stratosphere. In this case the photochemical lifetime is several orders of magnitude larger than the characteristic time for transport processes. Then we can treat methane as in photochemical equilibrium, which means that we can neglect the change of methane concentration due to photochemical processes. The fact that methane variability is mainly determined by transport makes it a good tracer of atmospheric masses in the stratosphere.

Example 8.8 Consider the reaction of nitric oxide and ozone,

$$NO + O_3 \rightarrow NO_2 + O_2. \tag{8.47}$$

Assuming that this reaction is the lone mechanism of NO depletion, find the lifetime of NO at temperature 250 K (typical of $z = 30$ km in the atmosphere).

Answer: The change of NO concentration due to this reaction can be described by the equation:

$$\frac{d}{dt}[NO] = -k_1[NO][O_3] \tag{8.48}$$

where $k_1 = 1.8 \times 10^{-12} \exp(-1370/T)$ cm^3 s^{-1}. The concentration of O_3 can be considered as a constant since it is much larger than that of NO. If $[NO]_0$ is the concentration of NO at the initial time, then we have

$$[NO] = [NO]_0\, e^{-k_1[O_3]t}. \tag{8.49}$$

The lifetime $t_c = 1/k_1[O_3]$. With $k_1 \approx 7.5 \times 10^{-15}$ cm^3 s^{-1} at $T = 250$ K and concentration of O_3 equal to 3×10^{12} cm^{-3} at 30 km, we obtain $t_c \approx 40$ s.   □

8.5 Equilibrium constant

Consider a generic two-bodies-to-two-bodies reaction:

$$A + B \rightarrow C + D. \tag{8.50}$$

As the reaction proceeds the products of this reaction (C and D) become sufficiently dense in number that they will begin to react and form A and B through the reverse reaction:

$$C + D \rightarrow A + B. \tag{8.51}$$

When eventually the rate of the forward reaction (8.50) is the same as the rate of the reverse reaction (8.51), the system is in chemical equilibrium. We can combine (8.50) and (8.51) and write

$$A + B \rightleftharpoons C + D. \tag{8.52}$$

If we were trying to find the rate for which the reaction (8.52) proceeds from the point of view of the individual gas molecules, we would say the rate of increase of the concentration of C is given by

$$\frac{d}{dt}[C] = k_{ab}[A][B] - k_{cd}[C][D]$$

$$= \frac{d}{dt}[D]$$

$$= -\frac{d}{dt}[A]$$

$$= -\frac{d}{dt}[B] \tag{8.53}$$

where k_{ab} and k_{cd} are the reaction rate coefficients for forward and reverse reactions respectively. The equation simply states that the rate of buildup of C is the sum of the rates of reactive collisions of A and B minus the reverse process in which C and D react. The first term must be proportional to the respective number densities and similarly for the second (loss) term. Since for every creation of a C molecule there must be a B molecule, these rates of formation must be equal to each other and equal to the negative of the rates of formation of the A and B molecules.

In equilibrium the rates of change of the species are zero. This means

$$\frac{d}{dt}[C] = 0 \quad \text{equilibrium} \tag{8.54}$$

or

$$K = \frac{[C][D]}{[A][B]} = \frac{k_{ab}}{k_{cd}}. \tag{8.55}$$

The constant K is called the *equilibrium constant* for the reaction. If K is known, we can determine the ratios of concentrations of the product and reactant gases in equilibrium.

For the case of a general reaction

$$a\,A + b\,B \rightarrow c\,C + d\,D, \tag{8.56}$$

the equilibrium constant is (as derived in physical chemistry texts):

$$\boxed{K = \frac{[C]^c [D]^d}{[A]^a [B]^b}} \tag{8.57}$$

The rule can be generalized to cases where more than two species are on each side of the equation.

It is seen from (8.57) that when the inverse reaction rate is very small, which means that products dominate over reactants, K is large. Small K means there will be a relatively large concentration of the reactant species.

The equilibrium constant depends on collision dynamics and in principle should have a strong temperature dependence, since the intermolecular relative velocity will be an important factor in the rearrangements. To find the temperature dependence of the equilibrium constant, let us write the reaction rate coefficients for the forward and reverse reactions (8.52). Using the Arrhenius equation (8.59) the reaction rate coefficient for the forward reaction, k_1, is

$$k_1(T) = A_1 \, \exp\left(\frac{-E_{\text{act}}}{R*T}\right). \tag{8.58}$$

The reaction rate coefficient for the reverse reaction, k_2, is

$$k_2(T) = A_2 \, \exp\left(\frac{-E_{\text{act}} + \Delta\overline{H}^\circ}{R*T}\right). \tag{8.59}$$

The equilibrium constant

$$\boxed{K(T) = \frac{k_2}{k_1} \propto \exp\left(\frac{-\Delta\overline{H}^\circ}{R*T}\right)} \tag{8.60}$$

Hence, knowing the standard enthalpy associated with a reaction provides information about the equilibrium concentrations of the species.

Example 8.9 Consider the recombination of OH and O:

$$OH + O \rightarrow O_2 + H. \tag{8.61}$$

Write the expression for the equilibrium constant. By using formula (8.60) find out whether the products will dominate over reactants at high or low temperatures.

Answer: In order for the reaction to proceed, an OH molecule has to bump into an O atom in the same tiny volume. The rate of growth term must be proportional to [OH][O]. The equilibrium constant is

$$K = \frac{[O_2][H]}{[OH][O]}. \tag{8.62}$$

To find the temperature at which the forward reaction will be favored we need to find the standard enthalpy:

$$\Delta \overline{H}^\circ = \Delta \overline{H}^\circ(O_2) + \Delta \overline{H}^\circ(H) - \Delta \overline{H}^\circ(OH) - \Delta \overline{H}^\circ(O)$$

$$= (0 + 217.97 - 38.96 - 249.17) \text{ kJ mol}^{-1}$$

$$= -70.2 \text{ kJ mol}^{-1}. \tag{8.63}$$

Since $\Delta \overline{H}^\circ$ is negative, from (8.60) we have that the equilibrium constant K is larger with lower temperature. This explains why the products of reaction (8.61) will dominate over reactants at the lower temperatures. □

The next step is to find the connection between the equilibrium constant and the Gibbs energy. The equilibrium of the chemical reaction implies that at a given temperature there exist partial pressures of the gases involved in the reaction with which the rate of the forward reaction is equal to the rate of the reverse reaction. Let us express the equilibrium constant in terms of partial pressures of each constituent. Using the Ideal Gas Law for the molar concentrations we obtain:

$$K = \frac{(p_C/R^*T)^c (p_D/R^*T)^d}{(p_A/R^*T)^a (p_B/R^*T)^b} = \frac{p_C^c p_D^d}{p_A^a p_B^b}(R^*T)^\Lambda, \tag{8.64}$$

where the p_i are partial pressures and $\Lambda = (a+b) - (c+d)$. We can rewrite (8.64) in the form:

$$K = K_p(R^*T)^\Lambda, \tag{8.65}$$

where

$$\boxed{K_p = \frac{p_C^c p_D^d}{p_A^a p_B^b}} \quad \text{[equilibrium constant for ideal gases].} \tag{8.66}$$

K_p is used as an equilibrium constant for chemical reactions involving species in the gaseous state.

For a reversible transformation at constant temperature the change of Gibbs energy is (see (4.97))

$$dG = V dp. \tag{8.67}$$

For 1 mol of ideal gas

$$d\overline{G} = \frac{R^*T}{p}dp. \tag{8.68}$$

(Note that the molar Gibbs energy $\overline{G}$ is also called the chemical potential.) By integrating this equation from the standard pressure level of 1 atm to some arbitrary pressure level p holding T constant we obtain

$$\overline{G} - \overline{G}_T^\circ = R^*T \ln p, \tag{8.69}$$

where p is in atm. In this formula $\overline{G}$ is the Gibbs energy at pressure p and temperature T and $\overline{G}_T^\circ$ is the Gibbs energy at 1 atm and temperature T.

Using (8.69) we can write the change of Gibbs energy for the general chemical reaction

$$\Delta\overline{G} = [c\Delta\overline{G}(C) + d\Delta\overline{G}(D)] - [a\Delta\overline{G}(A) + b\Delta\overline{G}(B)] \tag{8.70}$$

in the form

$$\Delta\overline{G} = [c\Delta\overline{G}_T^\circ(C) + d\Delta\overline{G}_T^\circ(D) - a\Delta\overline{G}_T^\circ(A) - b\Delta\overline{G}_T^\circ(B)]$$
$$+ cR^*T \ln p_C + dR^*T \ln p_D - aR^*T \ln p_A - bR^*T \ln p_B$$
$$= \Delta\overline{G}_T^\circ + R^*T \ln \frac{(p_C)^c (p_D)^d}{(p_A)^a (p_B)^b}. \tag{8.71}$$

The argument of the logarithmic function in the last formula is the equilibrium constant K_p (see (8.66)). Therefore,

$$\Delta\overline{G} = \Delta\overline{G}_T^\circ + R^*T \ln K_p. \tag{8.72}$$

At equilibrium $\Delta\overline{G} = 0$, and the change of Gibbs energy at a pressure of 1 atm and arbitrary temperature T, $\Delta\overline{G}_T^\circ$, is related to the equilibrium constant at pressure p and temperature T, K_p, by the simple relation:

$$\Delta\overline{G}_T^\circ = -R^*T \ln K_p. \tag{8.73}$$

The equation (8.73) tells us that if $\Delta\overline{G}_T^\circ$ is positive, K_p should be less than unity, which means that at equilibrium the concentrations of the reactants will exceed those of the products. If, on the other hand, $\Delta\overline{G}_T^\circ$ is negative and, moreover, is large, then K_p is large and the products will dominatein the equilibrium.

For standard conditions at a pressure of 1 atm and a temperature of 25 °C we obtain, in joules,

$$\Delta \overline{G}^{\circ} = -R^* \times T_0 \times \ln K_p = -2478.9 \times \ln K_p \qquad (8.74)$$

The change in Gibbs energy ΔG is especially useful because it simultaneously takes into account both the First and Second Laws of Thermodynamics. It does so in such a way that if the temperature and pressure are held constant (and they often are in atmospheric problems) we have a function which can be applied much more broadly.

Example 8.10 Consider the reaction of recombination of NO_2 and O_3:

$$NO_2 + O_3 \rightleftharpoons NO_3 + O_2. \qquad (8.75)$$

Do the reactants or products dominate for the forward reaction at 1 atm and 25 °C? *Answer*: We have to find the change of the Gibbs energy for this reaction.

$$\Delta \overline{G}^{\circ} = (115.9 + 0 - 51.3 - 163.2) \, kJ \, mol^{-1} = -98.6 \, kJ \, mol^{-1}. \qquad (8.76)$$

From (8.74) we obtain $K_p = 1.9 \times 10^{17}$. With such a large value of K_p the products will dominate for the forward reaction at equilibrium.   □

8.6 Solutions

Chemistry refresher A *solution* is a homogeneous mixture of several components. Consider a two-component solution. One component has a mole fraction η_w (the ratio of the number of moles of a component to the total number of moles), and the other component has mole fraction η. The component with a greater mole fraction, let it be η_w, is called the *solvent*. The solvent determines the state of matter of the solution (gas, liquid or solid). The component with the smaller mole fraction, η, is called the *solute*. In Chapter 5 we considered a cloud droplet as an example of a solution with water as a solvent and the salt as a solute. A solution of a salt in a solvent such as water is *saturated* when the rates of dissolving and crystallization are equal. In this case there could be some substance in the crystalline form present in the composite system. For example, there might be a salt crystal inside a cloud droplet. The amount of dissolved material (solute) in the saturated solution is called its *solubility*, which might depend strongly on temperature and weakly on pressure. We are interested in the effect of the dissolved solute on the vapor pressure of the solvent.

Raoult's Law Consider a solution that is in chemical equilibrium. The vapor pressure of each component of the solution is approximately

$$p_i = p_i^{\text{pure}} \, \eta_i \qquad (8.77)$$

where p_i^{pure} is the vapor pressure of a pure ith component and η_i is the mole fraction of an ith component of the solution. The solution is called *ideal* when both solvent and solute obey Raoult's Law. Raoult's Law applies when the components of the solution are present in high concentrations. We used Raoult's Law when we considered the equilibrium vapor pressure over a droplet containing dissolved electrolytes.

For solutions at low concentrations the vapor pressure of the solute obeys *Henry's Law*. According to Henry's Law, the vapor pressure of a solute, p, is a product of the mole fraction of the solute, η, and an empirical tabulated constant, K_H, expressed in units of pressure:

$$p = K_H \, \eta. \qquad (8.78)$$

Generally, the value of K_H increases with increasing temperature. Thus, at the same pressure the mole fraction of a solute decreases with increasing temperature.

When the atmospheric pressure decreases, the partial pressure of a gas decreases, and the molar solubility of a gas decreases. For example, high in the mountains the atmospheric pressure is low; as a result the solubility of oxygen in human blood decreases, which can cause respiration problems. At the opposite end, the higher the pressure, the higher the solubility of gases. You might say the gas is "squeezed" into the solution.

Example 8.11 Calculate the molar solubility of nitrogen dissolved in 1 l of water at 25 °C and atmospheric pressure of 1 atm. Henry's Law constant for nitrogen at 25 °C is 8.68×10^9 Pa. The percentage by volume of N_2 in dry air is 78.1.
Answer: The partial pressure of N_2 at 1 atm is $p_{N_2} = 0.781 \times 1\,\text{atm} = 7.91 \times 10^4$ Pa. From Henry's Law $\eta_{N_2} = p_{N_2}/K_H = (7.91 \times 10^4\,\text{Pa})/(8.68 \times 10^9\,\text{Pa}) = 9.1 \times 10^{-6}$. The mole fraction of nitrogen $\eta_{N_2} = v_{N_2}/(v_{N_2} + v_{H_2O}) \approx v_{N_2}/v_{H_2O}$ since $v_{N_2} \ll v_{H_2O}$. The number of moles of H_2O in 1 l is $(1000/18)$ mol. Then, $v_{N_2} = (9.1 \times 10^{-6} \times 1000/18)$ mol $= 5.05 \times 10^{-4}$ mol. The molar solubility of nitrogen is 5.05×10^{-4} mol l^{-1}.   □

Example 8.12 Calculate the molar solubility of CO_2 in moles per liter dissolved in water at 25 °C and CO_2 pressure of 2.4 atm (pressure used to carbonate soda). Henry's Law constant for CO_2 at 25 °C is 1.67×10^8 Pa.
Answer: 2.4 atm $= 2.43 \times 10^5$ Pa. The mole fraction of CO_2 according to Henry's Law is $\eta_{CO_2} = p_{CO_2}/K_H = (2.43 \times 10^5\,\text{Pa})/(1.67 \times 10^8\,\text{Pa})$. Since there is $(1000/18)$ mol of H_2O in 1 l, the molar solubility of CO_2 is $(2.43 \times 10^5\,\text{Pa} \times 1000/18\,\text{mol l}^{-1})/(1.67 \times 10^8\,\text{Pa}) = 8.1 \times 10^{-2}$ mol l^{-1}. When one opens a bottle of soda, the pressure decreases; as a result the solubility of CO_2 decreases, and the

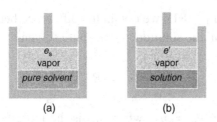

Figure 8.7 Notation. (a) Pure solvent in equilibrium with its vapor at pressure e_s. (b) Solution in equilibrium with the solvent's vapor at pressure e'.

bubbles of CO_2 emerge from the solution. At higher temperature the solubility of CO_2 decreases since Henry's constant increases (soda from a fridge sparkles less than soda held at room temperature).   □

8.6.1 Molar Gibbs energy of an ideal solution

In this section we will find out how the Gibbs energy of a pure solvent changes when a small amount of solute is added. We will consider liquids that are at equilibrium with their vapors. This means that the Gibbs energy of the vapor is equal to the Gibbs energy of the liquid. Let us denote the equilibrium vapor pressure over a pure solvent as e_s, and the equilibrium vapor pressure over the solution as e' (see Figure 8.7). From (8.69) the molar Gibbs energy (the Gibbs energy per mole) of vapor at pressure e_s is

$$\overline{G} = \overline{G}_T^\circ + R^* T \ln e_s \tag{8.79}$$

where e_s is in atm, $\overline{G}$ is the molar Gibbs energy[6] at pressure p and temperature T and $\overline{G}_T^\circ$ is the molar Gibbs energy at 1 atm and temperature T. Since at equilibrium the molar Gibbs energy of a liquid is equal to that of the vapor, $\overline{G}_{\text{vapor}} = \overline{G}_{\text{liquid}}$, the Gibbs energy of a pure solvent, denoted as $\overline{G}_w$, is

$$\overline{G}_w = \overline{G}_T^\circ + R^* T \ln e_s. \tag{8.80}$$

When a solute is added, the molar Gibbs energy of the solvent, $\overline{G}'$, which is at equilibrium with its vapor at pressure e', is

$$\overline{G}' = \overline{G}_T^\circ + R^* T \ln e'. \tag{8.81}$$

---

[6] The molar Gibbs energy is often called the chemical potential.

214 *Thermochemistry*

Subtracting (8.80) from (8.81), we obtain the difference between the molar Gibbs energy of the solvent in the solution and the pure solvent:

$$\overline{G'} - \overline{G}_w = R^*T \ln \frac{e'}{e_s}. \tag{8.82}$$

From Raoult's Law, $e'/e_s = \eta_w$, where η_w is the mole fraction of solvent (e.g. water) in the solution. Then,

$$\overline{G'} - \overline{G}_w = R^*T \ln \eta_w. \tag{8.83}$$

We will consider a dilute solution, when the mole fraction of the solute, η, is much smaller than that of the solvent, $\eta \ll \eta_w$. Since $\eta + \eta_w = 1$, we rewrite (8.83) as

$$\overline{G'} - \overline{G}_w = R^*T \ln (1 - \eta). \tag{8.84}$$

Taking into account that for $\eta \ll 1$, the logarithmic function can be written as $\ln(1 - \eta) \approx -\eta$, we obtain

$$\overline{G}_w - \overline{G'} = R^*T \, \eta. \tag{8.85}$$

We will use (8.85) in the next section to find the temperature at which the solution boils and freezes.

8.6.2 The elevation of the boiling point and the lowering of the freezing point of a solution

When some solute is dissolved in a pure solvent, the boiling and freezing points of the solution are not the same as for the pure solvent. We will show that the change is proportional to the amount of solute. We will consider nonvolatile solutes (for example, a salt). In this case the vapor of the solute is a pure gas.

Example 8.13 Show that the addition of a dissolved solute in a solution will elevate the boiling point compared to the boiling point of pure solvent.
Answer: The boiling point is the temperature at which the saturated vapor pressure of a liquid is the same as the atmospheric pressure. We consider two cases.

Case 1: equilibrium between a pure solvent and its vapor (see Figure 8.8a). The pure solvent boils at temperature T_0. The vapor pressure is 1 atm. At equilibrium, the molar Gibbs energies of the pure solvent, denoted below as $\overline{G}_w$, and its vapor, denoted as $\overline{G}_v$, are equal to each other:

$$\overline{G}_v(T_0, p) = \overline{G}_w(T_0, p). \tag{8.86}$$

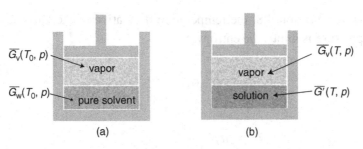

Figure 8.8 Notation. (a) Molar Gibbs energy for the vapor over the pure solvent, $\overline{G}_v(T_0, p)$, and molar Gibbs energy for the pure solvent, $\overline{G}_w(T_0, p)$. (b) Molar Gibbs energy of solvent vapor over the solution, $\overline{G}_v(T, p)$, and molar Gibbs energy of the solvent, $\overline{G}'(T, p)$.

Case 2: there is an amount η of nonvolatile solute in the solvent. The solution is at equilibrium with the solvent's vapor (see Figure 8.8b). The vapor pressure at the boiling point is 1 atm, the same as in case 1, but the boiling point differs. Assume that the solution boils at a temperature $T = T_0 + \Delta T$ where $\Delta T \ll T_0$. From (8.85) the molar Gibbs energy of the solvent in the solution is $\overline{G}' = \overline{G}_w - R^*T\eta$. At equilibrium, the molar Gibbs energy of the solvent in the solution, $\overline{G}'$, is equal to the molar Gibbs energy of the vapor, $\overline{G}_v$:

$$\overline{G}_v(T_0 + \Delta T, p) = \overline{G}'(T_0 + \Delta T, p) \tag{8.87}$$

or

$$\overline{G}_v(T_0 + \Delta T, p) = \overline{G}_w(T_0 + \Delta T, p) - R^*T\eta. \tag{8.88}$$

Since $\Delta T \ll T_0$ we can expand both sides of (8.88) in Taylor's series retaining only the linear term:

$$\overline{G}_v(T_0 + \Delta T, p) = \overline{G}_v(T_0, p) + \frac{\partial \overline{G}_v}{\partial T}\bigg|_{(T_0, p)} \Delta T, \tag{8.89}$$

$$\overline{G}_w(T_0 + \Delta T, p) = \overline{G}_w(T_0, p) + \frac{\partial \overline{G}_w}{\partial T}\bigg|_{(T_0, p)} \Delta T. \tag{8.90}$$

Substituting these expansions in (8.88) and taking into account (8.86), we obtain

$$R^*T_0\eta = \left(\frac{\partial \overline{G}_w}{\partial T}\bigg|_{(T_0, p)} - \frac{\partial \overline{G}_v}{\partial T}\bigg|_{(T_0, p)} \right) \Delta T. \tag{8.91}$$

Here we also replaced $T\eta$ by $T_0\eta$ since η has a small value proportional to ΔT.

As shown in Section 4.8, the temperature derivative of the Gibbs energy at a constant pressure is minus the entropy. So,

$$\frac{\partial \overline{G}_w}{\partial T}\bigg|_{(T_0,p)} = -\overline{s}_w \tag{8.92}$$

and

$$\frac{\partial \overline{G}_v}{\partial T}\bigg|_{(T_0,p)} = -\overline{s}_v \tag{8.93}$$

where $\overline{s}_w$ is the entropy of 1 mol of the solvent and $\overline{s}_v$ is the entropy of 1 mol of the solvent's vapor. Then, from (8.91) we obtain

$$R^*T_0\eta = (\overline{s}_v - \overline{s}_w)\Delta T \tag{8.94}$$

and

$$\Delta T = \frac{R^*T_0\eta}{\overline{s}_v - \overline{s}_w}. \tag{8.95}$$

Multiplying and dividing the right-hand side of the last equation by T_0 and taking into account that $(\overline{s}_v - \overline{s}_w)T_0$ is the amount of heat required to evaporate 1 mol of the solvent at the boiling point, in other words $(\overline{s}_v - \overline{s}_w)T_0 = \Delta_{vap}\overline{H}^\circ$, we obtain the final formula for the elevation of the boiling point ΔT:

$$\boxed{\Delta T = \frac{R^*T_0^2\eta}{\Delta_{vap}\overline{H}^\circ}} \tag{8.96}$$

Since ΔT is positive, the boiling point of the solution is higher than that of the pure solvent. The change in the boiling point is proportional to the amount of solute η.   ☐

Example 8.14 What is the change in the boiling temperature of 1 l of water with 15 g of NaCl dissolved in it?
Answer: The molecular weight of NaCl is 58.44 g mol^{-1}. Substituting in (8.96) $\eta = (15/58.44)/(1000/18)$, $R^* = 8.31$ J K^{-1} mol^{-1}, $T_0 = 373$ K, $\Delta_{vap}\overline{H}^\circ = 40.656$ kJ mol^{-1}, we obtain $\Delta T = 0.13$ K.   ☐

Example 8.15 Calculate the change in the freezing point of the solution.
Answer: We assume that only pure solvent is frozen, while the solute remains in the solution. Then the calculation of the freezing point of the solution is similar to the calculation of the boiling point except we have fusion instead of vaporization. $\Delta_{fus}\overline{H}^\circ$ is the heat released when 1 mol of the solvent is frozen at a temperature T_0;

this heat is negative. Substituting $\Delta_{\text{fus}}\overline{H}^{\circ}$ in (8.96), we find that the freezing point of a solution is lower than that of the pure solvent. The difference ΔT is:

$$\Delta T = -\frac{R^* T_0^2 \eta}{|\Delta_{\text{fus}}\overline{H}^{\circ}|}. \tag{8.97}$$

The decrease in the freezing point is proportional to the amount of solute. □

Example 8.16 What is the change in the freezing temperature of 1 l of water with 15 g of NaCl dissolved in it?
Answer: $|\Delta_{\text{fus}}\overline{H}^{\circ}|$ for water is 6.008 kJ mol^{-1}, $T_0 = 273$ K, $\eta = (15/58.44)/(1000/18)$, $R^* = 8.31$ J K^{-1} mol^{-1}. According to (8.97) the change in freezing temperature $\Delta T = -0.48$ K. □

Notes
There are excellent books on physical chemistry, for example Atkins (1994). The book by Houston (2001) gives a readable account of the kinetic theory of gases and reaction kinetics. The book by Hobbs (2000) on physical chemistry for the atmospheric sciences is at about the same level as this book and it delves much more into the subject of reactions in the atmosphere. A more thorough discussion of chemical equilibrium is contained in Denbigh (1981). The book on atmospheric physics and chemistry by Seinfeld and Pandis (1988) is the most comprehensive. Books on the more general subject of atmospheric chemistry include those by Brimblecombe (1986) and Warneck (1999). A comprehensive book on the chemistry of the middle atmosphere is by Brasseur and Solomon (2005). Finlayson-Pitts and Pitts (2000) cover both the upper and lower atmosphere and include many useful tables.

Notation and abbreviations for Chapter 8
c speed of light in vacuum (m s^{-1})
$\Delta\overline{G}^{\circ}$ change in Gibbs energy per mole, at standard conditions (kJ mol^{-1})
$\Delta\overline{H}^{\circ}$ change in enthalpy per mole, at standard conditions (kJ mol^{-1})
E_{act} activation energy (J)
E energy (J)
η mixing ratio
f frequency of electromagnetic wave (Hz)
F energy flux (J m^{-2} s^{-1})
$F_\lambda(z)$ flux of electromagnetic energy at wavelength λ, elevation z (J m^{-2} s^{-1} m^{-1})

g, l, aq   gaseous, liquid, aqueous phase
h Planck's constant (J s)
$\overline{H}^{\circ}$ standard enthalpy (kJ mol^{-1})
J photodissociation coefficient (s^{-1})
k reaction coefficient
K, K_p equilibrium constant
K_H Henry's Law constant
λ wavelength of an electromagnetic wave (m)
n_0 number density of photons (photons m^{-3})
$N(z)$ concentration of absorbers at level z (molecules m^{-3})
R^* universal gas constant (J mol^{-1} K^{-1})
$\sigma(\lambda)$ absorption cross-section (m^2)
t_c lifetime (s)
$t_{1/2}$ half life (s)
τ optical depth (dimensionless)
Θ zenith angle
v^* threshold velocity to exceed an activation barrier (m s^{-1})
$[X]$ concentration of X (molecules cm^{-3})
$[X]_0$ concentration of X at $t = 0$

Problems

8.1 Compute the standard enthalpy of reaction for the following reactions:

$$CO + O_2 \rightarrow CO_2 + O$$

$$O + O_3 \rightarrow 2O_2.$$

Are these reactions exothermic or endothermic?

8.2 Calculate the standard Gibbs energy for the reaction:

$$NO_3 + H_2O \rightarrow HNO_3 + OH.$$

Can this reaction proceed spontaneously?

8.3 Consider the reaction

$$2NO \rightarrow N_2 + O_2.$$

Determine whether high or low temperatures are favorable for the forward and reverse reactions. (*Hint*: Use formula (8.60).)

8.4 The reaction of HNO$_3$ dissociation is

$$HNO_3 + hf \rightarrow OH + NO_2.$$

By examination of the standard enthalpy find out what photon energy is required for this reaction to proceed (use Table 8.1).

8.5 Calculate the equilibrium constant at 298 K for the reaction

$$NO + O_3 \rightarrow NO_2 + O_2.$$

Are the reactants or products favored for this reaction at equilibrium at 298 K? Do you need to raise or lower the temperature to have more products at the equilibrium?

8.6 Find the expression for the equilibrium constant for the reaction

$$A + B \rightleftharpoons AB,$$

in terms of reaction coefficients, if the coefficient for the forward reaction is k_1, and the coefficient for the reverse reaction is k_2.

8.7 Estimate the molar solubility of oxygen in water at 25 °C and 1 atm. Henry's constant for oxygen is 4.4×10^9 Pa. The percentage by volume of O_2 in dry air is 20.95%.

8.8 Estimate by how much the amount of nitrogen in a diver's blood will change when the diver is rising from a depth of 80 m. An adult human male has an average blood volume of about 5 l. Use Henry's Law constant for nitrogen at 25 °C, $K_H = 8.68 \times 10^9$ Pa. (Fast rising, i.e., a rapid decrease in pressure and thus nitrogen solubility, can cause the formation of nitrogen bubbles in the bloodstream which often leads to death.)

8.9 Show that the difference between the vapor pressure of a solution at a given temperature T and the vapor pressure of a pure solvent at the same temperature is

$$\Delta p = -\frac{R^* T \eta}{v_v - v}$$

where η is a mole fraction of a nonvolatile solute, v is the volume of 1 mol of a solvent, v_v is the volume of 1 mol of vapor. *Hint*: Suggest that the solution is in equilibrium with the vapor at pressure $e' = e_s + \Delta p$ where e_s is the equilibrium vapor pressure for the pure solvent and Δp is small. The fact that Δp is small allows you to expand the Gibbs energy in Taylor's series and retain only the linear term (similar to the calculation of the boiling point of a solution).

9

The thermodynamic equation

In this chapter we derive two of the fundamental equations of atmospheric science, the *equation of continuity* and the *thermodynamic equation*. The equation of continuity expresses the conservation of mass in the form of a partial differential equation, the form needed to implement it in numerical simulations or forecasts. The thermodynamic equation expresses the combined First and Second Laws of Thermodynamics into a similar form. But before we come to these important formulas we need some experience with scalar and vector fields. Much of the chapter is concerned with elementary vector analysis which should have been covered in the prerequisite calculus course. Hence, some students can skip over the review sections, but we advise all students to refresh their memories. All this machinery is to prepare for the next step in an education in atmospheric sciences: *dynamics*.

Vector refresher A three-dimensional vector, denoted in boldface, $\mathbf{a}$, is a mathematical object which has both length and direction. In two-space it can be represented by an arrow as in Figure 9.1. It takes three numbers to represent a 3-vector, two angles and a length, $(\theta, \phi; |\mathbf{a}|)$. Alternatively, it can be represented by its three components along the three Cartesian coordinate axes, (a_x, a_y, a_z). Note that the vector is an abstract object in space independent of the choice of coordinate system, but the three numbers needed to specify it may individually depend on the coordinate system chosen by the analyst to describe the vector. For example, it is conventional in meteorology to set up a Cartesian coordinate system with the origin at a point on the Earth's surface, the x-axis increasing in the eastward direction and the y-axis increasing in the northward direction.

Multiplication by a scalar Let α be a scalar, i.e., a number which is independent of our choice of coordinate system, $\mathbf{a}$ be a vector. Then multiplication of a vector by a scalar is written as $\mathbf{b} = \alpha \mathbf{a}$. The direction of $\mathbf{b}$ is the same as that of $\mathbf{a}$ and the length is $|\alpha||\mathbf{a}|$. In other words the two angles designating the direction of $\mathbf{b}$ and $\mathbf{a}$ are the

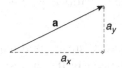

Figure 9.1 The vector **a** in two dimensions.

same, but the length is stretched. If α is negative, the direction of the vector is reversed. In the case of Cartesian component representation, all three components are multiplied by α: $(\alpha a_x, \alpha a_y, \alpha a_z)$.

Adding vectors The sum of two vectors is a vector $\mathbf{c} = \mathbf{a} + \mathbf{b}$ whose components are defined by adding the Cartesian components of the addends:

$$(c_x, c_y, c_z) = (a_x + b_x, a_y + b_y, a_z + b_z). \tag{9.1}$$

Clearly $\mathbf{a} + \mathbf{b} = \mathbf{b} + \mathbf{a}$. The sum of the two vectors can also be understood geometrically by joining the tail of one vector to the head of the other and the line joining the first tail to the second head is the vector sum.

Cartesian unit vectors A unit vector is a vector which has unit length. It is very convenient and common to use unit vectors which point along the three Cartesian axes:

$$\mathbf{i} = (1, 0, 0) \tag{9.2}$$

$$\mathbf{j} = (0, 1, 0) \tag{9.3}$$

$$\mathbf{k} = (0, 0, 1). \tag{9.4}$$

This notation allows us to *expand* a vector into its Cartesian components

$$\mathbf{a} = a_x \mathbf{i} + a_y \mathbf{j} + a_z \mathbf{k}, \quad \text{or alternatively } (a_x, a_y, a_z). \tag{9.5}$$

The length of **a** is given by

$$|\mathbf{a}| = \sqrt{a_x^2 + a_y^2 + a_z^2}. \tag{9.6}$$

Note that the length of a vector is independent of the coordinate system chosen.

Wind in meteorology is usually denoted by $\mathbf{v} = u\mathbf{i} + v\mathbf{j} + w\mathbf{k}$. Thus a wind blowing to the east is $u\mathbf{i}$; if $u > 0$ it is called a *westerly*. A wind blowing to the north is $v\mathbf{j}$; if $v > 0$ it is called *southerly*. Rising air (a vertical wind) is denoted $w\mathbf{k}$. Note that the components are referred to a Cartesian coordinate system whose origin is at the surface of the Earth at a fixed location. Wind *speed* is given by $\sqrt{u^2 + v^2 + w^2}$.

Example 9.1 Find the length of the vector $\mathbf{a} = 2\mathbf{i} + 3\mathbf{j} - 4\mathbf{k}$.
Answer: $|\mathbf{a}| = \sqrt{2^2 + 3^2 + (-4)^2} = \sqrt{29}.$ ☐

Example 9.2 Find the sum of the vectors $\mathbf{a} = 2\mathbf{i} + 3\mathbf{j} - 4\mathbf{k}, \mathbf{b} = -4\mathbf{i} + 3\mathbf{j} + 2\mathbf{k}$.
Answer: $\mathbf{a} + \mathbf{b} = -2\mathbf{i} + 6\mathbf{j} - 2\mathbf{k}$. $\qquad\qquad\qquad\qquad\qquad\qquad\qquad$ □

It is sometimes useful to use a curvilinear coordinate system such as **cylindrical coordinates** to describe phenomena in the atmosphere. We use the unit vectors pointing along the cylindrical coordinate directions (dependent on the point): $\hat{\mathbf{r}}, \hat{\mathbf{n}}_\theta, \hat{\mathbf{k}}$. A vector $\mathbf{a}$ can be written as:

$$\mathbf{a} = a_r\,\hat{\mathbf{r}} + a_\theta\hat{\mathbf{n}}_\theta + a_z\,\hat{\mathbf{k}}. \qquad (9.7)$$

Example 9.3 A purely cyclonic wind blows counterclockwise (northern hemisphere) about a center of action. At a distance r from the center the wind can be denoted as $v_\theta(r, \theta)\hat{\mathbf{n}}_\theta$. $\qquad\qquad\qquad\qquad\qquad\qquad\qquad\qquad\qquad\qquad$ □

Example 9.4 A point on a rotating disk with axis at the disk's center has velocity $r\omega\hat{\mathbf{n}}_\theta$, where ω is angular velocity in rad s^{-1} and r is the axis of rotation. $\qquad$ □

Vector refresher: dot product The *dot* or *scalar product* of two vectors results in a scalar. Even though the components of the two vectors forming the scalar product depend on the choice of coordinate system, the scalar product does not:

$$\mathbf{a} \cdot \mathbf{b} = |\mathbf{a}||\mathbf{b}|\cos(\theta_{\mathbf{a},\mathbf{b}}) \qquad (9.8)$$

where $\theta_{\mathbf{a},\mathbf{b}}$ is the angle between the two vectors as shown in Figure 9.2. Note that since $\cos\theta_{\mathbf{a},\mathbf{b}}$ is an even function of its argument, the order of the vectors in the product makes no difference: $\mathbf{a} \cdot \mathbf{b} = \mathbf{b} \cdot \mathbf{a}$.

An alternative way to define the dot product $\mathbf{a} \cdot \mathbf{b}$ is

$$\mathbf{a} \cdot \mathbf{b} = a_x b_x + a_y b_y + a_z b_z. \qquad (9.9)$$

It can be shown that the two definitions are equivalent. Using the second definition,

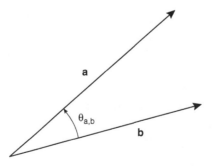

Figure 9.2 Illustration of the vectors $\mathbf{a}$ and $\mathbf{b}$ in a dot product.

---

we can show that $\mathbf{c} \cdot (\mathbf{a} + \mathbf{b}) = \mathbf{c} \cdot \mathbf{a} + \mathbf{c} \cdot \mathbf{b}$.

It is also clear that $\mathbf{i} \cdot \mathbf{i} = 1, \mathbf{i} \cdot \mathbf{j} = 0$, etc.

Example 9.5 Find $\mathbf{a} \cdot \mathbf{b}$ where $\mathbf{a} = 2\mathbf{i} + 3\mathbf{j} - 4\mathbf{k}, \mathbf{b} = -4\mathbf{i} + 3\mathbf{j} + 2\mathbf{k}$.

Answer: $\mathbf{a} \cdot \mathbf{b} = (2) \times (-4) + 3 \times 3 + (-4) \times (2) = -7$. □

Example 9.6 What is the angle between $\mathbf{a}$ and $\mathbf{b}$?

Answer: $\cos \theta = \mathbf{a} \cdot \mathbf{b}/|\mathbf{a}||\mathbf{b}| = -7/29$; arccos $(-7/29) = 1.815$. □

Cross product The *vector* or *cross product* is indicated by the notation

$$\mathbf{c} = \mathbf{a} \times \mathbf{b}. \tag{9.10}$$

The vector $\mathbf{c}$ is perpendicular to the plane defined by $\mathbf{a}$ and $\mathbf{b}$ (see Figure 9.3). The right hand rule is used to determine its direction (point the right index finger along the first vector in the product, then sweep it toward the second vector's direction; the thumb points in the direction of the vector product). Its length is given by

$$|\mathbf{c}| = |\mathbf{a}||\mathbf{b}| \sin(\theta_{\mathbf{a},\mathbf{b}}). \tag{9.11}$$

By the right hand rule we see:

$$\mathbf{a} \times \mathbf{b} = -\mathbf{b} \times \mathbf{a}. \tag{9.12}$$

A useful form for the cross product $\mathbf{a} \times \mathbf{b}$ is

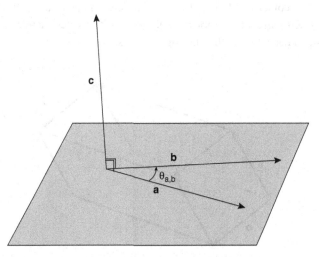

Figure 9.3 Illustration of the vectors $\mathbf{c} = \mathbf{a} \times \mathbf{b}$ in a cross product. $\mathbf{c}$ is perpendicular to the plane formed by $\mathbf{a}$ and $\mathbf{b}$.

$$\mathbf{a} \times \mathbf{b} = \begin{vmatrix} \mathbf{i} & \mathbf{j} & \mathbf{k} \\ a_x & a_y & a_z \\ b_x & b_y & b_z \end{vmatrix}. \tag{9.13}$$

Math refresher: 3×3 determinant Recall that it can be expanded along any row or column. In the representation of the cross product:

$$\begin{vmatrix} \mathbf{i} & \mathbf{j} & \mathbf{k} \\ a_x & a_y & a_z \\ b_x & b_y & b_z \end{vmatrix} = \mathbf{i}(a_y b_z - a_z b_y) - \mathbf{j}(a_x b_z - a_z b_x) + \mathbf{k}(a_x b_y - a_y b_x).$$

Recall the alternating signs of the unit vectors as one expands along the top row. Note that if two rows (vectors) of a determinant are proportional, the determinant vanishes. This is simply the statement that $\sin(\theta_{\mathbf{a},\mathbf{b}}) = 0$.

Vector refresher: the box product This is defined by:

$$\mathbf{a} \cdot (\mathbf{b} \times \mathbf{c}) = \mathbf{c} \cdot (\mathbf{a} \times \mathbf{b}) = \mathbf{b} \cdot (\mathbf{c} \times \mathbf{a}). \tag{9.14}$$

A useful form for the box product is

$$\mathbf{c} \cdot (\mathbf{a} \times \mathbf{b}) = \begin{vmatrix} c_x & c_y & c_z \\ a_x & a_y & a_z \\ b_x & b_y & b_z \end{vmatrix}. \tag{9.15}$$

The three vectors $\mathbf{a}$, $\mathbf{b}$ and $\mathbf{c}$ form a rectangular parallelepiped. The box product is the volume of that geometrical figure. The rules of determinants can be useful here. For example, cyclic permutation of the rows leads to an equivalent determinant. Interchanging adjacent rows flips the sign. It is useful to know that the three vectors

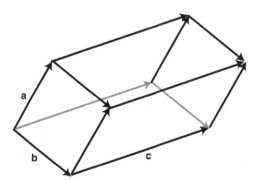

Figure 9.4 Sketch of the parallelepiped formed by the vectors $\mathbf{a}$, $\mathbf{b}$ and $\mathbf{c}$. The magnitude of the box product $\mathbf{a} \cdot \mathbf{b} \times \mathbf{c}$ is the volume of the parallelepiped.

of a box product can be identified as the three edges of a solid parallelepiped. The box product is the volume of the parallelepiped (see Figure 9.4). Note that

$$\mathbf{k} \cdot (\mathbf{i} \times \mathbf{j}) = 1. \tag{9.16}$$

9.1 Scalar and vector fields

A scalar field is a function defined on the three-dimensional space coordinates and possibly along the time axis. An example is the temperature field $T(x, y, z; t) \equiv T(\mathbf{r}, t)$, where the position vector $\mathbf{r}$ is defined by

$$\mathbf{r} \equiv x\mathbf{i} + y\mathbf{j} + z\mathbf{k} \tag{9.17}$$

and $\mathbf{i}, \mathbf{j}, \mathbf{k}$ are unit vectors pointing along the x, y and z axes (see Figure 9.5). A small increment in $\mathbf{r}$ is denoted as [1]

$$\mathbf{dr} = dx\,\mathbf{i} + dy\,\mathbf{j} + dz\,\mathbf{k}. \tag{9.18}$$

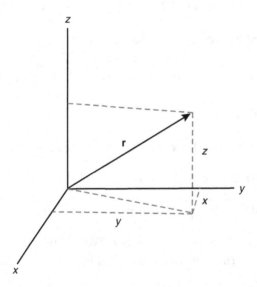

Figure 9.5 Schematic diagram of a position vector $\mathbf{r}$ whose components are x, y and z.

---

[1] Here we replace the small values $\Delta x, \delta x$, etc., with infinitesimals dx, etc., with the approximate sign $\approx$ replaced by the equality sign $=$. This means that in this notation second-order quantities such as $(dx)^2$ are neglected (set to zero) when additive to first-order terms. While this *operational* shortcut might cause some to cringe, it should not disturb the flow of our story.

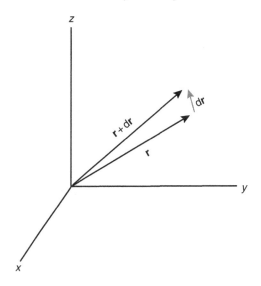

Figure 9.6 Schematic diagram of a position vector **r** and an increment of it d**r**.

Consider the estimation of the temperature field at the point **r** + d**r** (Figure 9.6), given that we know its value at the point **r**, namely $T(\mathbf{r})$:

$$T(\mathbf{r} + d\mathbf{r}) = T(x + dx, y + dy, z + dz).$$

We may use the first two terms of the Taylor expansion:

$$T(x + dx, y + dy, z + dz) = T(x, y, z) + \frac{\partial T}{\partial x} dx + \frac{\partial T}{\partial y} dy + \frac{\partial T}{\partial z} dz. \quad (9.19)$$

We can also write this as a dot product:

$$T(\mathbf{r} + d\mathbf{r}) = T(\mathbf{r}) + d\mathbf{r} \cdot \nabla T(\mathbf{r}). \quad (9.20)$$

After substituting $dT = T(\mathbf{r} + d\mathbf{r}) - T(\mathbf{r})$, we obtain

$$\boxed{dT = d\mathbf{r} \cdot \nabla T} \quad \text{[differential of a scalar field].} \quad (9.21)$$

The vector $\nabla T(\mathbf{r})$ is called the *gradient of T*. We will use the modern notation ∇T to denote the gradient (in some older texts it is denoted **grad** *T*).

$$\boxed{\nabla T = \frac{\partial T}{\partial x} \mathbf{i} + \frac{\partial T}{\partial y} \mathbf{j} + \frac{\partial T}{\partial z} \mathbf{k}} \quad \text{[gradient of a scalar field].} \quad (9.22)$$

The gradient is a *vector field*. At each point in space **r** it has an associated length and direction.

If you want to know the rate of change of the field in a particular direction, say along the direction defined by the unit vector, $\mathbf{n}$, it can be found by defining[2] the vector increment $d\mathbf{r}$ to be $\mathbf{n}\,ds$ where ds is an infinitesimal distance and $\mathbf{n}$ defines the direction along which the increment is to be taken. Using (9.21) we can write

$$\frac{dT}{ds} = \mathbf{n} \cdot \nabla T \quad \text{[derivative in the direction } \mathbf{n}\text{].} \qquad (9.23)$$

This derivative taken along the direction of the specified unit vector $\mathbf{n}$ is called the *directional derivative*, and is often given the notation $\partial T/\partial n$ as the rate of change of T along a certain direction, defined by the unit vector $\mathbf{n}$. The conventional notation for the directional derivative is:

$$\frac{\partial T}{\partial n} = \mathbf{n} \cdot \nabla T \quad \text{[directional derivative].} \qquad (9.24)$$

If $\mathbf{n}$ lies in the tangent plane to an isothermal (still thinking of the scalar field as temperature) surface, the directional derivative vanishes since there is no change in any direction lying in this plane. This means that the component (projection) of the gradient vector tangent to the isothermal surface vanishes. The gradient vector is perpendicular to isothermal surfaces (in general so-called *level surfaces*). This can be seen for a fixed gradient vector ∇T. Just vary the unit vector in all directions. The lengths of $\mathbf{n}$ and ∇T are fixed, so the maximum occurs when the angle between $\mathbf{n}$ and ∇T is zero ($\cos \theta_{\mathbf{n},\nabla T} = 1$), in other words when $\mathbf{n}$ is parallel to ∇T.

Example 9.7 Consider the field

$$T(x, y) = T_0 \cos 2\pi x \cos \pi y. \qquad (9.25)$$

Find the gradient as a function of x and y.
Answer:

$$\nabla T(x, y) = -\pi T_0 \left(2 \sin 2\pi x \cos \pi y \mathbf{i} + \cos 2\pi x \sin \pi y \mathbf{j}\right). \qquad (9.26)$$

See the contour map in Figure 9.7. ☐

Example 9.8 Find the directional derivative of the field in the last example in the direction $\mathbf{n} = (1/\sqrt{2})(\mathbf{i} + \mathbf{j})$ (this is a unit vector in the x–y plane directed $45°$ above the x-axis).
Answer: Take the dot product of $\mathbf{n}$ with the gradient:

$$\mathbf{n} \cdot \nabla T = \frac{-\pi T_0}{\sqrt{2}} \left(2 \sin 2\pi x \cos \pi y + \cos 2\pi x \sin \pi y\right). \qquad ☐$$

---

[2] Remember that the reader has the power to choose $d\mathbf{r}$, its tiny length and direction.

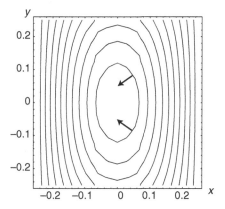

Figure 9.7 Contour map of a field $T(\mathbf{r}) = \cos 2\pi x \cos \pi y$ showing constant T lines. Arrows indicate direction of the gradient vector evaluated at the points where the arrows originate.

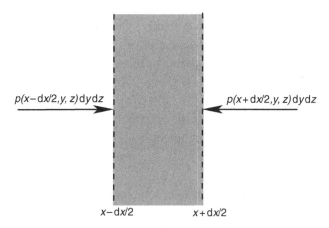

Figure 9.8 A small rectangular parallelepiped of side widths dx, dy, dz, indicating the pressure forces on the sides perpendicular to the x-axis.

9.2 Pressure gradient force

The gradient of the pressure field $\nabla p(\mathbf{r}, t)$ is very important in meteorology. Consider a parcel of air contained in the rectangular parallelepiped $dx\,dy\,dz$, and whose center is located at the point $\mathbf{r}$. This volume element is embedded in a surrounding pressure field $p(\mathbf{r}) = p(x, y, z)$. Let us compute the x component of the net force on the volume element. As indicated in Figure 9.8, the left-most face experiences a force due to the external field to the right:

$$F_x^{\text{left face}} = p(x - \tfrac{1}{2}\,dx, y, z)\,dy\,dz, \qquad (9.27)$$

the right-most face experiences a force to the left

$$F_x^{\text{right face}} = -p(x + \tfrac{1}{2}\,dx, y, z)\,dy\,dz. \tag{9.28}$$

The net force on the volume element is

$$
\begin{aligned}
dF_x^{\text{net}} &= -\left(p(x + \tfrac{1}{2}\,dx, y, z) - p(x - \tfrac{1}{2}\,dx, y, z)\right)\,dy\,dz \\
&= -\frac{\partial p}{\partial x}\,dx\,dy\,dz = -\frac{\partial p}{\partial x}\,dV
\end{aligned}
\tag{9.29}
$$

where dV is the volume of the infinitesimal material element. Newton's Second Law (force is mass times acceleration) tells us that

$$(dM)a_x = dF_x^{\text{net}} \tag{9.30}$$

where a_x is the x component of acceleration and dM is the mass contained in the parcel. We can divide each side by dV and obtain

$$\rho a_x = -\frac{\partial p}{\partial x} \tag{9.31}$$

where ρ is the density of the air in the parcel. Put in more conventional form we have:

$$a_x = -\frac{1}{\rho}\frac{\partial p}{\partial x}. \tag{9.32}$$

If we evaluate the y and z components in a similar fashion we can summarize the result in vector form

$$\boxed{\mathbf{a} = -\frac{1}{\rho}\nabla p} \quad \text{[pressure gradient force/mass].} \tag{9.33}$$

The force per unit mass **a** as given here is called the *pressure gradient force*. We have encountered its vertical component earlier in establishing the hydrostatic equation. Above the atmospheric boundary layer its horizontal components are very nearly balanced by the Coriolis force in midlatitudes (called *geostrophic balance*).

Example 9.9: horizontal acceleration of a parcel in the tropics Suppose a parcel whose density is $0.7\,\text{kg m}^{-3}$ is embedded in a field of pressure with a gradient $10\,\text{hPa}$ over $1000\,\text{km}$. What is the acceleration of the parcel (ignoring the Coriolis force) and what is its increase in speed from rest in passing over $1000\,\text{km}$? *Answer:* The acceleration is toward low pressure and is given by $a = 1.43 \times 10^{-3}\,\text{m s}^{-2}$. $v = \sqrt{2ax} = 53\ \text{m s}^{-1}$. ☐

9.3 Surface integrals and flux

Consider the vector field of velocity of a fluid, $\mathbf{v}(\mathbf{r})$, at an instant of time (disregard time dependence for now). The fluid is moving locally along line segments tangential to the local velocity vector $\mathbf{v}(\mathbf{r})$. The trajectories of individual parcels follow the flow lines in steady flow ($\partial \mathbf{v}/\partial t = 0$) but differ in unsteady flow ($\partial \mathbf{v}/\partial t \neq 0$). An imaginary surface is placed in the fluid flow (say a penetrable screen) as indicated in Figure 9.9. Let an element of the surface be denoted by $d\mathbf{S}$ where dS is the magnitude of the (vector) area element and the direction of $d\mathbf{S}$ is the local perpendicular to the surface area element. In a closed surface, by convention, the vector points outwards; otherwise, it has to be specified according to context.

First imagine a parallel flow that is uniform over its cross-section in a pipe in the x direction. Then $\mathbf{v}(\mathbf{r}) = v_0 \, \mathbf{i}$. Suppose the surface $\mathbf{S}$ is a perpendicular cross-section of the pipe. Then $\mathbf{S} = S\mathbf{i}$. How much mass passes through S per unit time? First consider the "front" of fluid passing through S at time t. At time $t + dt$ the front advances by a distance $v_0 dt$. The amount of volume swept out by this front in the time dt is just $v_0 S \, dt = \mathbf{v} \cdot \mathbf{S} \, dt$ as in Figure 9.9. If the density of the fluid is ρ, we can convert this to a mass flux

$$d\mathcal{M} = \rho \mathbf{v} \cdot \mathbf{S} \, dt \tag{9.34}$$

or

$$\frac{d\mathcal{M}}{dt} = \rho \mathbf{v} \cdot \mathbf{S}. \tag{9.35}$$

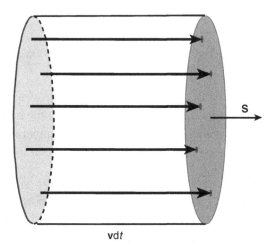

Figure 9.9 Advance of a material surface through a cylindrical pipe during the time dt. The flow is taken to be of uniform velocity $\mathbf{v}$ over the cross-section.

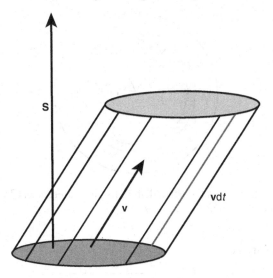

Figure 9.10 Illustration of a flow through a surface at an angle to the normal. The volume of fluid swept out in the time dt is $|\mathbf{S}||\mathbf{v}|dt\cos\theta_{\mathbf{S},\mathbf{v}} = \mathbf{S}\cdot\mathbf{v}\,dt$.

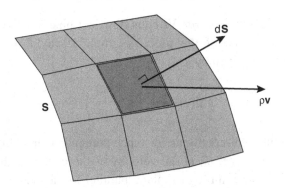

Figure 9.11 Portion of a surface $\mathbf{S}$ depicting an area element on a surface $d\mathbf{S}$ through which fluid of density ρ is passing at velocity $\mathbf{v}$.

If the surface were tilted with respect to the y–z plane the amount of mass per unit time passing through it would be the same (see Figure 9.10). The only thing that matters is the projection of $\mathbf{S}$ onto the velocity $\mathbf{v}$. If the surface were curved we would have to generalize to (see Figure 9.11)

$$\frac{d\mathcal{M}}{dt} = \iint_S \rho\mathbf{v}\cdot d\mathbf{S}. \tag{9.36}$$

The amount of *mass flux* passing through the area element $d\mathbf{S}$ is

$$\boxed{\text{mass flux} = \rho\mathbf{v}\cdot d\mathbf{S}}\quad\text{[flux of mass through an area element]}. \tag{9.37}$$

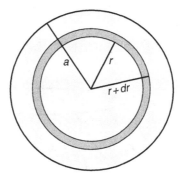

Figure 9.12 The annulus is the shaded area. Its circumference is $2\pi r$ and its width is dr. The area of the annulus is $2\pi r\, dr$.

The mass flux density is

$$\boxed{\text{mass flux density} = \rho\mathbf{v}\cdot\mathbf{n}} \qquad (9.38)$$

where $\mathbf{n}$ is a unit vector parallel to the infinitesimal area vector $d\mathbf{S}$ (i.e., $\mathbf{n} = d\mathbf{S}/dS$).

Example 9.10: fluid flow in a pipe Viscous flow in a pipe is slower near the walls than along the centerline of the pipe. A simple steady flow solution for an incompressible fluid is given by

$$u(r) = u_0\left(1 - \frac{r^2}{a^2}\right) \qquad (9.39)$$

where r is the cylindrical-coordinate distance perpendicular to the centerline and a is the radius of the pipe. u_0 is the velocity at the center of the cross-section. What is the flux of mass through a plane parallel to the centerline of the pipe? *Answer:* First form an annulus (a ring) (see Figure 9.12) in the cross-section. The area of the ring is $2\pi r\, dr$. The mass flux through the ring is $\rho u(r)2\pi r\, dr$. The total flux is

$$F = \int_0^a \rho u(r)2\pi r\, dr = \frac{1}{2}\pi\rho u_0 a^2. \qquad (9.40)$$

$\square$

9.4 Conduction of heat

The Fourier Law of heat conduction states that the amount of heat (enthalpy) flowing across a unit perpendicular area (the vector enthalpy flux $\mathbf{q}$) is proportional to the gradient of the temperature field, ∇T. The Fourier Law works well in solids since the transfer is from one molecule to its neighbor in a medium where there is no relative macroscopic motion from one location in the solid to another.

In liquids or gases the story can be much more complicated because these macroscopic motions are permitted. Buoyancy for example might cause differential forces moving lighter (usually warmer) material upwards leaving the more dense fluid behind. This results in a net transfer of heat upwards in the medium at the macroscopic level. While the actual transfer of heat takes place from one infinitesimal element to another via molecular collisions (a relatively slow process when considered at macroscopic scales), the macroscopic motions can move heat around much more rapidly than pure molecular transfers at the smaller scale from one infinitesimal element to another building up to the macroscopic scale.

The transfer of heat by winds or currents is called *thermal advection*. In atmospheric applications the transfer is dominated by advection by large eddies (fluctuating or irregular departures typical in turbulence from the larger scale flows). For example, in the morning boundary layer where turbulence is common, the air at the surface which has been heated by the rising sun can be buoyed in parcels to heights of a kilometer or two (where its rise might be limited by increased stability at those levels). The eddies necessarily bring warm air in a parcel into contact with cooler air at the same level with an ensuing large thermal gradient at boundaries separating warmer and cooler parcels and ultimately enthalpy is transferred at the molecular level:

$$\mathbf{q} = -\kappa_H \nabla T(\mathbf{r}) \tag{9.41}$$

where $\mathbf{q}$ is a heat flux density and κ_H is a coefficient known as the thermal conductivity (it varies from one substance to another and can be found in tables). Heat flows from warm toward cool regions in the direction opposite the gradient vector. The amount of heat transferred by molecular processes per unit time flowing through a surface S (flux), F_S, is

$$F_S = \iint_S \mathbf{q} \cdot d\mathbf{S} = - \iint_S \kappa_H (\nabla T(\mathbf{r})) \cdot d\mathbf{S}. \tag{9.42}$$

Example 9.11 Steady heat flows along a rod with circular cross-section (area A) and length L with its left and right ends attached to reservoirs of temperatures T_L and T_R. Let $x = 0$ at the left end and $x = L$ at the right end of the rod. The flux of heat $F(x)$ at the point x is

$$F(x) = -A\kappa_H \frac{dT}{dx}. \tag{9.43}$$

But the flux of heat must be constant at any point along the rod otherwise heat energy would accumulate at some point. Then $F(x) = F_0$. We can now integrate

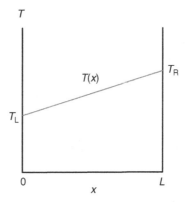

Figure 9.13 Temperature dependence along a rod of length L, held at temperature T_L at $x = 0$ and T_R at $x = L$.

the last equation from 0 to L to find:

$$F_0 L = -A\kappa_H(T_R - T_L).\tag{9.44}$$

This allows us to evaluate F_0:

$$F_0 = A\kappa_H\frac{T_L - T_R}{L}\tag{9.45}$$

and we find the x dependence of $T(x)$ by integrating from 0 to x:

$$T(x) = T_L + \frac{F_0}{A\kappa_H}x = T_L + \left(\frac{T_R - T_L}{L}\right)x.\tag{9.46}$$

See Figure 9.13. It is interesting that the curve does not depend on κ_H or A. □

9.5 Two-dimensional divergence

We begin our study of divergence in two dimensions (in the x–y plane). We are examining an important property of a vector field such as the two-dimensional velocity $\mathbf{V}(x, y)$. Is there more fluid flowing out of a small fixed area in the plane than is coming in? The two-dimensional divergence is relevant in meteorological applications. For example, at a box (fixed in space) surrounding a low pressure area at the surface, air spirals in counterclockwise in the northern hemisphere (cyclonically) towards the center of the low. In the x–y plane (the surface) there is a net flow of air into the box. What happens to it? (After all, mass is conserved.) The answer is it goes up in the z direction. When air goes up we know what happens: it

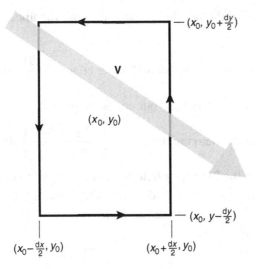

Figure 9.14 A two-dimensional rectangular box centered at the point (x_0, y_0) in a constant z plane illustrating the flux issuing from the box.

rains (assuming moisture is available, etc.). Hence we care about the net flow into or out of fixed horizontal boxes at different levels in the atmosphere.[3]

There is a mathematical way of expressing the net flow into or out of a two-dimensional (2D) box. Let us focus our question on a small region in space consisting of a rectangle of sides dx and dy whose center is at (x_0, y_0). We can evaluate the flux passing through each of the four edges and add them up to find out whether there is a net flux issuing *from* the box. To obtain the flux coming through the right hand edge (see Figure 9.14) we need to take

$$\mathbf{V}(x_0 + \tfrac{1}{2} dx, y) \cdot \mathbf{i} dy = V_x(x_0 + \tfrac{1}{2} dx, y) \, dy. \tag{9.47}$$

The flux going out of the left vertical edge is similar:

$$\mathbf{V}(x_0 - \tfrac{1}{2} dx, y) \cdot (-\mathbf{i}) dy = -V_x(x_0 - \tfrac{1}{2} dx, y) dy. \tag{9.48}$$

When we add these left and right edge flux contributions together we obtain

$$\text{net flux out the left and right edges} = \frac{\partial V_x}{\partial x} \, dx \, dy. \tag{9.49}$$

---

[3] Note that the air in the box might have simply become more dense during the net inflow, but we implicitly made the approximation that the air is very nearly incompressible.

It is easy to see that the net flux passing through the upper and lower edges is just

$$\text{net flux out from the upper and lower edges} = \frac{\partial V_y}{\partial y} \, dx \, dy. \tag{9.50}$$

If we add up the fluxes from all four edges we obtain:

$$\text{total flux leaving the box} = \left(\frac{\partial V_x}{\partial x} + \frac{\partial V_y}{\partial y} \right) dx \, dy. \tag{9.51}$$

The *divergence* of the 2D vector field $\mathbf{V}(x, y)$ is defined to be the limit as the box shrinks to zero of the emerging flux divided by the area of the box. We can express it in Cartesian coordinates using our results from above:

$$\boxed{\text{div}_2(\mathbf{V}) = \frac{\partial V_x}{\partial x} + \frac{\partial V_y}{\partial y}} \quad \text{[2D divergence]} \tag{9.52}$$

where we have employed the subscript 2 to make it clear that we are dealing with two dimensions only. But the definition holds more generally for any infinitesimal loop (rectangular, parallelogram, circle, etc.) around the tiny shrinking area:

$$\text{div}_2(\mathbf{V}) = \lim_{\Delta A \to 0} \frac{1}{\Delta A} \oint \mathbf{V} \cdot (\mathbf{k} \times d\mathbf{r})$$

$$= \lim_{\Delta A \to 0} \frac{1}{\Delta A} \oint (\mathbf{V} \times \mathbf{k}) \cdot d\mathbf{r}. \tag{9.53}$$

In the last step we made use of the rule for triple vector box products: $\mathbf{a} \cdot (\mathbf{b} \times \mathbf{c}) = (\mathbf{a} \times \mathbf{b}) \cdot \mathbf{c}$.

Example 9.12: easterly flow rate increasing Suppose $\mathbf{v}(x) = \lambda x \mathbf{i}, \lambda > 0$. The divergence of $\mathbf{v}(x)$ is $\text{div}_2(\mathbf{v}) = \lambda$. This is a divergent flow since more fluid is leaving a tiny box (fixed in space anywhere in the flow) on the east side per unit time than is entering on the west side (per unit box area and time). □

Example 9.13: divergence expressed in polar coordinates Following Figure 9.15 we use a loop around an area element in polar coordinates. The four sides are (1) $r \to r + dr, \theta$. (2) At the outer radius $r + dr$, let $\theta \to \theta + d\theta$. (3) Now decrease r: at $\theta + d\theta, r + dr \to r$. (4) Now back to the starting point, at $r, \theta + d\theta \to \theta$.

The two angle-changing sides yield for the emerging (radial) fluxes (steps 1 and 3):

$$\text{net radial fluxes} = V_r(r + dr, \theta)(r + dr)\, d\theta - V_r(r, \theta) r \, d\theta$$

$$= \frac{\partial(rV_r)}{\partial r} \, d\theta \, dr. \tag{9.54}$$

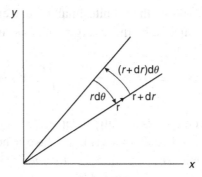

Figure 9.15 View in polar coordinates of the box described in Example 9.13.

Next consider the angular fluxes (steps 2 and 4):

$$\text{net angular fluxes} = -V_\theta(r, \theta)\,dr + V_\theta(r, \theta + d\theta)\,dr$$

$$= \frac{\partial V_\theta}{\partial \theta}\,dr\,d\theta. \tag{9.55}$$

The area of the little element is $r\,d\theta\,dr$. Upon dividing through by this area and taking the limit as it shrinks to zero we have:

$$\text{div}_2(\mathbf{V}) = \frac{1}{r}\frac{\partial(rV_r)}{\partial r} + \frac{1}{r}\frac{\partial V_\theta}{\partial \theta}. \tag{9.56}$$

$\square$

Example 9.14: rigidly rotating fluid This is similar to a rotating disk. The velocity field is $\mathbf{V} = v_\theta \hat{\mathbf{n}}_\theta = r\omega \hat{\mathbf{n}}_\theta$ where $\hat{\mathbf{n}}_\theta$ is a unit vector perpendicular to $\hat{\mathbf{r}}$ fixed at the point (r, θ) and pointing in the direction of increasing θ (see (9.7)). We can compute $\text{div}_2(\mathbf{V}) = 0$. Note that $V_r = 0$ in this case. This is a general result in the case of $\mathbf{V} = f(r)\hat{\mathbf{n}}_\theta$. $\square$

9.6 Three-dimensional divergence

The generalization to the three-dimensional divergence is straightforward. This time the tiny box in the plane becomes a 3D box fixed in the space. Consider a closed surface S. The velocity flux emanating from the enclosed volume is

$$\text{emerging flux} = \oiint_S \mathbf{v}(\mathbf{r}) \cdot d\mathbf{S}. \tag{9.57}$$

Now let the enclosed volume become very small. For the Cartesian coordinate case take it to be a rectangular parallelepiped whose sides are of lengths dx, dy, dz. Its volume is therefore the product, $(dV)_S = dx\,dy\,dz$ where the subscript S is used to indicate the surface surrounding the infinitesimal volume element. The *divergence* of the velocity field is defined to be the emerging flux per unit volume:

$$\text{div}(\mathbf{v}(\mathbf{r})) \equiv \lim_{(dV)_S \to 0} \frac{1}{(dV)_S} \oiint_S \mathbf{v}(\mathbf{r}') \cdot d\mathbf{S}'. \tag{9.58}$$

This definition of the divergence is actually independent of the shape of the volume for reasonably well-behaved functions $\mathbf{v}(\mathbf{r})$. We take it here to be a rectangular parallelepiped for convenience. Note that the divergence is a scalar field defined over the space whose points are designated by $\mathbf{r}$.

While this appears to be a useful concept, so far it seems to be a rather difficult thing to compute. Next we will find a convenient way to compute the divergence. In rectangular coordinates we take the surface to be the rectangular parallelepiped mentioned before.

The integration over the six faces of the box is so similar to the two-dimensional case that we need not repeat it here. The result is

$$\text{emerging flux} = \left(\frac{\partial v_x}{\partial x} + \frac{\partial v_y}{\partial y} + \frac{\partial v_z}{\partial z} \right) dx \, dy \, dz. \tag{9.59}$$

The divergence is then:

$$\boxed{\text{div}_3 \, \mathbf{v} = \frac{\partial v_x}{\partial x} + \frac{\partial v_y}{\partial y} + \frac{\partial v_z}{\partial z}} \tag{9.60}$$

From our earlier notation with the ∇ operator, we can write

$$\nabla \cdot \mathbf{v} = \frac{\partial v_x}{\partial x} + \frac{\partial v_y}{\partial y} + \frac{\partial v_z}{\partial z}$$

$$= \left(\mathbf{i} \frac{\partial}{\partial x} + \mathbf{j} \frac{\partial}{\partial y} + \mathbf{k} \frac{\partial}{\partial z} \right) \cdot \left(v_x \mathbf{i} + v_y \mathbf{j} + v_z \mathbf{k} \right). \tag{9.61}$$

Example 9.15: divergence of the product of scalar and vector We can find the divergence of a product of a scalar field and a vector field by expanding the individual terms into their Cartesian $[x, y, z]$ components. Let $G(\mathbf{r})$ be a scalar field and $\mathbf{a}(\mathbf{r})$ be a vector field:

$$\nabla \cdot (G(\mathbf{r})\mathbf{a}(\mathbf{r})) = G(\mathbf{r})\nabla \cdot \mathbf{a}(\mathbf{r}) + \mathbf{a}(\mathbf{r}) \cdot \nabla G(\mathbf{r}). \tag{9.62}$$

$\square$

9.7 Divergence theorem

Consider a macroscopic volume V surrounded by a closed surface S in which a vector field $\mathbf{A}(\mathbf{r})$ is defined. Let the flux out of the volume be

$$\text{flux out} = \oiint_S \mathbf{A} \cdot d\mathbf{S}. \tag{9.63}$$

Now subdivide the volume into boxes each one of which is small enough that we can use the approximation

$$(\nabla \cdot \mathbf{A})_i \approx \left(\frac{1}{V} \oiint \mathbf{A} \cdot d\mathbf{S} \right)_i. \tag{9.64}$$

Note that flux flowing out of the sides of one little box flows into the sides of its neighbor. If we form the sum

$$\sum_i (\nabla \cdot \mathbf{A})_i \, (dV)_i \approx \iiint_V \nabla \cdot \mathbf{A} \, dV. \tag{9.65}$$

But since the flux leaving the whole volume is just the algebraic sum of the fluxes leaving the tiny boxes

$$\text{flux out} \approx \sum_i \left(\oiint \mathbf{A} \cdot d\mathbf{S} \right)_i$$

$$= \oiint_S \mathbf{A} \cdot d\mathbf{S}. \tag{9.66}$$

We have at last the *divergence theorem*:

$$\oiint_S \mathbf{A} \cdot d\mathbf{S} = \iiint_V \nabla \cdot \mathbf{A} \, dV. \tag{9.67}$$

The divergence theorem is very useful in that it says we can apply the micro-definition essentially to macroscopic volumes. We simply integrate the divergence for micro-volumes up to obtain the flux issuing from the macroscopic volume. While the mathematical expression for the divergence in Cartesian coordinates is very useful for computation from analytical formulas, we often find that in meteorological applications the integral forms are easier to apply. For example in the 2D case we can integrate around a box on a weather map and divide by its area to obtain a good approximation to the divergence.

Example 9.16 A cylinder of radius R of air has density profile $\rho(z)$. It is rigidly rising at $w_0 \, \mathrm{m\,s^{-1}}$. What is the flux of mass passing through a level at $z = z_0$? *Answer*: Mass flux $= \rho(z) \cdot w_0 \cdot \pi R^2$. Just for fun take $R = 1 \, \mathrm{km}$; $\rho = \rho_0 e^{-z/H}$,

$\rho_0 = 1.2 \, \text{kg m}^{-3}$, $H = 10^4$ m; and finally $w_0 = 0.01 \, \text{m s}^{-1}$. Then at $z = 0$, we have $3.77 \times 10^4 \, \text{kg s}^{-1}$. At $z = H/2$ this becomes $2.29 \times 10^4 \, \text{kg s}^{-1}$. $\qquad\qquad\square$

9.8 Continuity equation

Consider the divergence of the product $\rho(\mathbf{r})\mathbf{v}(\mathbf{r})$. According to the definition of the divergence this is the flux of *mass* per unit time issuing from an infinitesimal volume per unit volume. If the box from which the mass is issuing is *fixed in space* (and there are no sources of mass inside), the mass inside the box has to be changing:

$$\text{loss of mass/time} = -\frac{d\mathcal{M}}{dt} = \oiint (\rho \mathbf{v}) \cdot d\mathbf{S} \qquad (9.68)$$

where $\mathcal{M}$ is the mass inside the fixed box. Dividing by the volume of the box and letting it shrink to zero, we obtain:

$$-\frac{\partial \rho}{\partial t} = \nabla \cdot (\rho \mathbf{v}). \qquad (9.69)$$

The minus sign takes into account that the flux out of the box represents a negative rate of change of mass in the box. Note that we used the *partial* derivative in the last formula because we are referring to a *fixed* position for our box. Rearranging we have the Eulerian form of the *equation of continuity*:

$$\boxed{\frac{\partial \rho}{\partial t} + \nabla \cdot (\rho \mathbf{v}) = 0} \quad \text{[Euler form of the equation of continuity].} \qquad (9.70)$$

By expanding the divergence of the product we can write it in another form:

$$\frac{\partial \rho}{\partial t} + \mathbf{v} \cdot \nabla \rho + \rho \nabla \cdot \mathbf{v} = 0. \qquad (9.71)$$

This last equation has a very special meaning if we regroup the first two terms

$$\boxed{\frac{D\rho}{Dt} = -\rho \nabla \cdot \mathbf{v}} \quad \text{[Lagrangian form of the equation of continuity]} \qquad (9.72)$$

where the differential operator

$$\boxed{\frac{D}{Dt} \equiv \frac{\partial}{\partial t} + \mathbf{v} \cdot \nabla} \quad \text{[material derivative]} \qquad (9.73)$$

is called the *material derivative*, and we will return to it in the next section.

An alternative and perhaps a more physical derivation of the continuity equation (in two dimensions for simplicity) can be conducted as follows. Consider a small

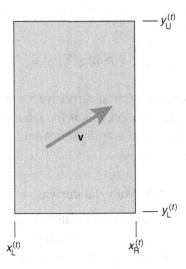

Figure 9.16 Schematic view of a box element in motion at velocity **v**. The box might extend its width or height during the motion, but mass must be conserved.

rectangular box with sides dx and dy, whose area is $dx\,dy$. The mass density of material in the box is ρ giving the mass in the box as $\mathcal{M} = \rho\,dx\,dy = \text{constant}$ since mass will be conserved along the path of the box. The box might be distorted due to the differential motions of the fluid (e.g., shear). We can write:

$$\frac{d\mathcal{M}}{dt} = \frac{d}{dt}\{\rho(x,y,t)\cdot(x_R(t) - x_L(t))(y_U(t) - y_L(t))\} \tag{9.74}$$

where $x_R(t)$ represents the location of the right hand edge of the box as it moves, $x_L(t)$ similarly represents the location of the left hand edge (see Figure 9.16). The same notation goes for the upper and lower edges of the box. Note that $x_R(t)$ might be moving at a different speed from $x_L(t)$ and therefore the box might be stretched or squeezed in that direction. As we take the derivative through the expression we obtain

$$\frac{d\mathcal{M}}{dt} = \frac{d\rho}{dt}\,dx\,dy + \rho\,du\,dy + \rho\,dx\,dv = 0 \tag{9.75}$$

where by du we mean $(dx/dt)(R) - (dx/dt)(L)$, and by dv we mean $(dy/dt)(U) - (dy/dt)(L)$. In the last equation we have to recognize that the derivative of ρ is *along* the motion, as was the derivative of the mass. After dividing through by the area $dx\,dy$ and taking the limit:

$$\frac{d\rho}{dt} + \rho\frac{du}{dx} + \rho\frac{dv}{dy} = 0. \tag{9.76}$$

Now with the fancy notation:

$$\frac{D\rho}{Dt} + \rho \, \mathrm{div}_2(\mathbf{V}) = 0. \tag{9.77}$$

In the last equation we used the material derivative for the rate of change of ρ since it is taken along the motion. Note that this is the same statement as in (9.72). The generalization to three dimensions is straightforward.

9.9 Material derivative

As a parcel moves, its temperature or some other property might change along the path of the parcel. On the other hand, some properties are conserved along the path, for example, the potential temperature in adiabatic flows. The rate of change of a field along the fluid's motion is an important point of view to take because many physical laws are most easily expressed in this form. For example, the rate of change of the momentum in reaction to an imposed force is to be taken along the path of the parcel (Newton's Second Law).

Take a small rectangular parallelepiped of dimension dx, dy, dz. Let the center of the figure move with the local velocity of the fluid in which it is embedded, $\mathbf{v}(\mathbf{r}, t)$. The velocity field $\mathbf{v}(\mathbf{r}, t)$ might be changing in both space and time. Consider the temperature $T(\mathbf{r}, t)$ as an example. If we know the value of the temperature at a certain point $(\mathbf{r}, t)$, say $T(\mathbf{r}, t)$, what can we say about its value at neighboring points in space-time? The total differential can be used to make an estimate. We can write the total differential for the temperature field as

$$\begin{aligned} dT &= \frac{\partial T}{\partial t} dt + \frac{\partial T}{\partial x} dx + \frac{\partial T}{\partial y} dy + \frac{\partial T}{\partial z} dz \\ &= \frac{\partial T}{\partial t} dt + d\mathbf{r} \cdot \nabla T. \end{aligned} \tag{9.78}$$

This is the change in the temperature due to a displacement from point $\mathbf{r}$ to $\mathbf{r} + d\mathbf{r}$ and from t to $t + dt$.

Now divide through by dt

$$\frac{dT}{dt} = \frac{\partial T}{\partial t} + \frac{d\mathbf{r}}{dt} \cdot \nabla T. \tag{9.79}$$

In general $d\mathbf{r}$ can be in any direction, but in this special case it should be the displacement $d\mathbf{r}$ that occurs due to the motion of the fluid during the time dt. In other words, $d\mathbf{r} = \mathbf{v}(\mathbf{r}, t) \, dt$ where $\mathbf{v}(\mathbf{r}, t)$ is the velocity of the fluid motion field evaluated at $(\mathbf{r}, t)$. After this crucial substitution we can write:

$$\frac{dT}{dt} = \frac{\partial T}{\partial t} + \mathbf{v} \cdot \nabla T. \tag{9.80}$$

Stated again, this is the explicit recognition that the small increment $\mathbf{dr}$ which is associated with the time interval dt is identified with the motion of the fluid, namely it is the spatial increment that is advanced by the fluid itself during the time dt. We again encounter the important combination of derivatives which we termed the *material derivative*,

$$\frac{D}{Dt} \equiv \frac{\partial}{\partial t} + \mathbf{v} \cdot \nabla. \tag{9.81}$$

The change of the temperature field along the motion, DT/Dt, is composed of two terms: the first term, $\partial T/\partial t$, represents the *local* change of the temperature in a certain volume fixed in space (a weather station for example); the second term, $\mathbf{v} \cdot \nabla T$, is due to the advection of hotter or colder air into this fixed volume by the wind.

9.10 Thermodynamic equation

Consider the heating of a moving parcel. We apply the First Law of Thermodynamics to an individual parcel. We can write

$$dH = dQ + V\,dp \tag{9.82}$$

where we have used dQ to indicate the amount of diabatic heating [4] of the parcel in time dt. The rate of change of the enthalpy of the parcel as it moves along its path is given by the material derivative:

$$\frac{DH}{Dt} = \frac{DQ}{Dt} + V\frac{Dp}{Dt} \tag{9.83}$$

where DQ/Dt is the rate of diabatic heating of the parcel ($\mathrm{J\,s^{-1}}$). Since the air can be treated as an ideal gas we may write $dH = c_p \mathcal{M}\,dT$:

$$c_p\mathcal{M}\frac{DT}{Dt} = \frac{DQ}{Dt} + V\frac{Dp}{Dt}. \tag{9.84}$$

This is an expression for the First Law of Thermodynamics for a fluid in motion. The *thermodynamic equation* follows from (9.84) by dividing through by $\mathcal{M}$, the

---

[4] The diabatic heating includes solar and terrestrial infrared absorption heating as well as heating due to condensation.

mass of the parcel (since mass is conserved along the path):

$$\boxed{c_p \frac{DT}{Dt} = \frac{DQ_M}{Dt} + \frac{1}{\rho}\frac{Dp}{Dt}} \quad \text{[thermodynamic equation]} \quad (9.85)$$

where DQ_M/Dt indicates the heating rate per unit mass of air; it is called the *diabatic* heating rate (per unit mass). The left-hand side of the thermodynamic equation is the heating rate of a parcel along its path. The heating rate is proportional to the rate of change of the temperature. The right-hand side of the equation tells us what contributes to that heating rate. First is the heating per unit mass of the parcel by such actions as radiative heating or condensation heating, collectively called DQ_M/Dt. The second term contributes to the rate of temperature change because of compression or expansion of the parcel as it moves along its path from one pressure to another.

Example 9.17 Suppose a 1 kg parcel of air moves horizontally along an isobaric surface and is heated by radiation by $4\,\mathrm{W\,kg^{-1}}$. What is the rate of change of temperature of the parcel?
Answer: Note that the derivative Dp/Dt vanishes because the parcel moves along an isobaric surface. Then we can find the rate of change of temperature from $(4\,\mathrm{W\,kg^{-1}})/(1004\,\mathrm{J\,kg^{-1}\,K^{-1}}) \approx 0.004\,\mathrm{K\,s^{-1}}$. □

Example 9.18 In the previous example suppose the parcel is moving eastward along the horizontal isobaric surface at a velocity of $3\,\mathrm{m\,s^{-1}}$ and that the eastward component of the gradient of temperature is given by $1.5\,\mathrm{K\,km^{-1}}$. What is the local (fixed position) rate of change of temperature?
Answer: We need to write the material derivative:

$$\frac{\partial T}{\partial t} + \mathbf{v} \cdot \nabla T = 0.004\,\mathrm{K\,s^{-1}}.$$

We seek the value of $\partial T/\partial t$. The horizontal velocity is $\mathbf{v} = 3\,\mathbf{i}\,(\mathrm{m\,s^{-1}})$ and $\nabla T = 0.0015\,\mathbf{i}\,(\mathrm{K\,m^{-1}})$. Hence, $\partial T/\partial t = (0.004 - 0.0045)\,\mathrm{K\,s^{-1}}$. □

Example 9.19 What is the rate of change (along its path) of temperature of a parcel of density $\rho = 1.0\,\mathrm{kg\,m^{-3}}$ which is rising such that $Dp/Dt = -0.3\,\mathrm{Pa\,s^{-1}}$? Assume the diabatic heating rate is zero.
Answer: We can calculate $DT/Dt = -0.3\,\mathrm{Pa\,s^{-1}}/(1.0\,\mathrm{kg\,m^{-3}} \times 1004\,\mathrm{J\,K^{-1}\,kg^{-1}}) \approx -0.0003\,\mathrm{K\,s^{-1}}$. □

Example 9.20 Continuing the previous example, what is the local (fixed position) rate of change of the temperature, assuming the environmental lapse rate is $10\,\mathrm{K\,km^{-1}}$?

Answer: First we need to estimate the vertical velocity, w. Assuming hydrostatic balance we can calculate

$$\frac{Dp}{Dt} \approx \frac{dp}{dz}\frac{dz}{dt}$$

and $dp/dz = -\rho g$; hence, $w \approx 0.03 \text{ m s}^{-1}$. Now proceed with

$$\frac{\partial T}{\partial z} = -0.010 \text{ K m}^{-1}.$$

Thus,

$$\mathbf{v} \cdot \nabla T = -0.0003 \text{ K s}^{-1}.$$

Finally, we can combine the two and find that $\partial T/\partial t = 0$. This is because we chose the adiabatic lapse rate as our vertical gradient. The reader should think of the consequences of a lapse rate that is greater or less than the adiabatic lapse rate. □

9.11 Potential temperature form

The thermodynamic equation can also be expressed in terms of the potential temperature by starting with the Second Law of Thermodynamics:

$$s = c_p \ln \theta, \quad \text{or in differential form} \quad T ds = c_p \frac{T}{\theta} d\theta = dQ_M. \qquad (9.86)$$

Using the material derivative we find

$$\boxed{c_p \frac{T}{\theta}\frac{D\theta}{Dt} = \frac{DQ_M}{Dt}} \quad \text{[potential temperature form of the thermodynamic equation].}$$

$$(9.87)$$

This form is equivalent to the form (9.85) and it shows more directly the influence of both the First and Second Laws of Thermodynamics. It is left as an exercise for the reader to show the equivalence of the two (Problem 9.13).

9.12 Contributions to DQ$_M$/Dt

DQ$_M$/Dt is a material derivative, which means the rate of change is taken along the motion of the infinitesimal volume element. Local heating of an infinitesimal volume element can be due to several sources. We list a few of them here.

Heating by conduction At the molecular level the heat exchange from one infinitesimal volume element to its neighbors with differing temperature is given by the divergence of the heat flux $\nabla \cdot \mathbf{h}(\mathbf{r}, t)$. This gives the cooling rate per unit volume of the moving element. To obtain the cooling rate per unit mass, one must divide by density $\rho(\mathbf{r}, t)$.

Heating by phase change As a moist parcel moves it might experience a temperature change and this could lead to condensation (or evaporation) onto droplets. The release of enthalpy is given by $L\,d\mathcal{M}_{vap} = \mathcal{M}_{air}L\,dw_s$, where $w_s(T)$ is the saturation mixing ratio of the volume element containing $\mathcal{M}_{air}$ of air. Hence, the heating rate per unit mass is simply $-L\,dw_s$ (note that dw_s is negative for condensation).

Heating by radiation In this case we have a certain radiation flux density of energy $\mathbf{F}(\mathbf{r}, t)$. The heating rate per unit volume is $-\nabla \cdot \mathbf{F}(\mathbf{r}, t)$. And the heating rate per unit mass is $-(1/\rho)\nabla \cdot \mathbf{F}(\mathbf{r}, t)$.

To summarize we have

$$\frac{DQ_{\mathcal{M}}}{Dt} = -\frac{1}{\rho}\nabla \cdot \mathbf{h}(\mathbf{r}) - L\frac{Dw_s}{Dt} - \frac{1}{\rho}\nabla \cdot \mathbf{F}(\mathbf{r}, t). \tag{9.88}$$

Sometimes a frictional heating term is included as well. Generally in applications such as numerical weather forecasting and climate modeling, the first term above is small compared to the others and is neglected.

Notes

This chapter is really an introduction to dynamics. Most dynamics books cover these subjects and many do so in more detail, see for example, Holton (1992).

Notation and abbreviations for Chapter 9

a, b	etc. arbitrary vectors
a, a_x	these are used for vector acceleration and its x component
a_x, a_y, a_z	the Cartesian components of vector **a**
A	a vector field
$\text{div}_2\mathbf{V}$	divergence of vector field **V** in two dimensions
D/Dt	material derivative
$DQ_{\mathcal{M}}/Dt$	the rate of heating of a moving parcel per unit mass $(\text{J}\,\text{s}^{-1}\,\text{kg}^{-1})$
$\nabla \cdot \mathbf{F}(\mathbf{r}, t)$	divergence of the vector field $\mathbf{F}(\mathbf{r}, t)$
$\nabla T = (\partial T/\partial x)\mathbf{i} + \cdots$	gradient of T
i, j, k	the Cartesian unit vectors
κ_H	thermal conductivity $(\text{J}\,\text{K}^{-1}\,\text{m}^{-1}\,\text{s}^{-1})$
L	enthalpy of evaporation (latent heat) $(\text{J}\,\text{kg}^{-1})$
n	unit vector
$\mathbf{n}_\theta$	unit vector in the theta direction (polar coordinates)
$\partial T/\partial n$	directional derivative of T
$\mathbf{r} = x\mathbf{i} + y\mathbf{j} + z\mathbf{k}$	position vector
$\hat{\mathbf{r}}$	unit vector in the **r** direction (polar coordinates)

ρ	density (kg m^{-3})
S, dS	surface, surface element vector
v	three-dimensional velocity vector field (m s^{-1})
V	upper case usually indicates two-dimensional velocity

Problems

9.1 By writing out the components, prove that

$$\frac{d}{dt}(\mathbf{a} \cdot \mathbf{b}) = \frac{d\mathbf{a}}{dt} \cdot \mathbf{b} + \mathbf{a} \cdot \frac{d\mathbf{b}}{dt}.$$

9.2 A horizontal wind is blowing from southwest to northeast at 30° north of east. Express it in terms of **i, j, k**.

9.3 Find a unit vector 45° north of east. Find a unit vector perpendicular to this one in the x–y plane.

9.4 The vectors $2\mathbf{i} + 3\mathbf{j}, 4\mathbf{i} - \mathbf{k}$ determine a plane. Find a *unit* vector perpendicular to this plane.

9.5 A parcel moves along a path:

$$y = y_0 + \alpha \sin \omega t$$

$$x = x_0 + \alpha \cos \omega t + \gamma t.$$

If $\gamma = 0$, what is the curve that describes the motion? Describe the motion when $\gamma > 0$. What is the position vector as a function of time? What is the velocity vector?

9.6 A certain pressure field varies spatially as

$$p(x, y, z) = p_0 + A \sin \frac{3}{R}x \, \cos \frac{\pi}{2L} y \, e^{-z/H}.$$

How many maxima are there at the surface ($z = 0$) and at the center of the channel ($y = 0$) as one goes around the circle (x goes from 0 to $2\pi R$)? What is the total differential of the pressure? What is the pressure gradient?

9.7 The wind has the form $\mathbf{v} = (v_0/r^\ell) \mathbf{n}_\theta$, $\mathbf{n}_\theta = -\sin \theta \mathbf{i} + \cos \theta \mathbf{j}$. What are u, v? What is the divergence of this wind, $\mathrm{div}_2(\mathbf{v})$?

9.8 A certain surface pressure field is given by

$$p(x, y) = a(x - x_0)^4 + b(y - y_0)^2, \quad a, b > 0.$$

Describe this pressure surface near (x_0, y_0). What is the pressure gradient as a function of x and y? What is the directional derivative in the direction 45° north of east?

9.9 The 500 hPa level is given by

$$Z_{500} = 5500 + 50 \sin(\phi - \omega t) \cos \frac{\pi y}{2L}, \quad -L \le y \le L$$

where the vertical heights are in meters and the longitudinal distances are in kilometers. L is about 30° of latitude. Suppose $\omega = 2\pi$ radians per month. Describe this disturbance.

9.10 Suppose the diabatic heating rate for a parcel is $10\,\mathrm{W\,kg^{-1}}$ and the wind is blowing at an angle of $60°$ north of east at $10\,\mathrm{m\,s^{-1}}$. Also suppose the temperature gradient is $0.1\,\mathrm{K\,km^{-1}}$ and directed due east. You may assume the horizontal gradient of pressure is zero. Find the rate of change of temperature along the motion and find the local rate of change of the temperature at a fixed point.

9.11 Sinking air. Suppose the atmospheric profile is isothermal. A parcel is descending at $2\,\mathrm{cm\,s^{-1}}$ and the diabatic heating rate is zero. What is the rate of temperature increase along the motion; at a fixed altitude? (Use reasonable values for parameters not given.)

9.12 Dry air is blowing from west to east at a speed of $u = 12\,\mathrm{m\,s^{-1}}$. The temperature gradient is $\nabla T = 2.0 \times 10^{-6}\,(\mathrm{K\,m^{-1}})\,\mathbf{i}$. The air is being heated diabatically at a rate $1.1 \times 10^{-5}\,\mathrm{K\,kg^{-1}\,s^{-1}}$. The air moves along an isobaric surface.

  (a) Write an expression for the rate of change of the temperature of the air along its path in terms of the heating rates (thermodynamic equation).

  (b) Based upon (a) evaluate the material derivative.

  (c) The air is passing over a fixed station. Find the rate of change of the temperature at the station.

9.13 Show that the two forms (9.85, 9.87) of the thermodynamic equation are equivalent by using $\theta = T\,(p/p_0)^{-\kappa}$.

Appendix A
Units and numerical values of constants

The units used in atmospheric science are the *Standard International* (SI) units. These are essentially the MKS units. The units of length are *meters*, abbreviated m; those for mass are the *kilogram*, abbreviated kg; and for time the unit is the *second*, abbreviated s. The units for velocity then are $m\,s^{-1}$. The unit of force is the *newton* ($1\,kg\,m\,s^{-2}$, abbreviated N). Tables A1–A3 show the SI units for some basic physical quantities commonly used in atmospheric science.

The unit of pressure is the *pascal* ($1\ N\,m^{-2} = 1$ Pa). This is of special importance in meteorology. In particular, atmospheric scientists like the *millibar* (abbreviated mb), but in keeping with SI units more and more meteorologists use the *hectopascal* (abbreviated hPa; $1\ hPa = 100\ Pa = 1$ mb). The *kilopascal* ($1\ kPa = 10\ hPa$) is the formal SI unit and some authors prefer it. *One atmosphere* (abbreviated 1 atm) of pressure is

$$1\ atm = 1.013\ bar$$
$$= 1013.25\ mb$$
$$= 1013.25\ hPa$$
$$= 101.325\ kPa$$
$$= 101325\ Pa$$
$$= 1.01325 \times 10^5\ Pa \qquad (A1)$$

and $1\ mb = 1$ hectopascal $= 100\ Pa$. Sometimes one encounters pressure in inches of mercury (in Hg) or millimeters of mercury (mm Hg); $1\ atm = 760.000\ mm\ Hg = 29.9213\ in\ Hg$.

The *dimensions* of a quantity such as density, ρ, can be constructed from the fundamental dimensions of length, mass, time and temperature, denoted by the letters L, M, T, θ respectively. The dimensions of density, indicated with square brackets $[\rho]$, are ML^{-3}. In the SI system its *units* are $kg\,m^{-3}$. Many quantities are pure numbers and have no dimension; examples include arguments of functions such as sine or log. The radian is really not a unit in the sense used here.

Table A1 *Useful numerical values*

Universal

gravitational constant (G)	$6.673 \times 10^{-11} \mathrm{N\,m^2\,kg^{-2}}$
universal gas constant (R^*)	$8.3145 \mathrm{\,J\,K^{-1}\,mol^{-1}}$
Avogadro's number (N_A) [gram mole]	6.022×10^{23} molecules $\mathrm{mol^{-1}}$
Boltzmann's constant (k_B)	$1.381 \times 10^{-23} \mathrm{\,J\,K^{-1}}$ molecule^{-1}
proton rest mass	$1.673 \times 10^{-27} \mathrm{\,kg}$
electron rest mass	$9.109 \times 10^{-31} \mathrm{\,kg}$

Planet Earth

equatorial radius	$6378 \mathrm{\,km}$
polar radius	$6357 \mathrm{\,km}$
mass of earth	$5.983 \times 10^{24} \mathrm{\,kg}$
rotation period (24 h)	$8.640 \times 10^4 \mathrm{\,s}$
acceleration of gravity (at about 45 °N)	$9.8067 \mathrm{\,m\,s^{-2}}$
solar constant	$1370 \mathrm{\,W\,m^{-2}}$

Dry air

gas constant (R_d)	$287.0 \mathrm{\,J\,K^{-1}\,kg^{-1}}$
molecular weight (M_d)	$28.97 \mathrm{\,g\,mol^{-1}}$
speed of sound at $0\,°\mathrm{C}$, 1000 hPa	$331.3 \mathrm{\,m\,s^{-1}}$
density at $0\,°\mathrm{C}$ and 1000 hPa	$1.276 \mathrm{\,kg\,m^{-3}}$
specific heat at constant pressure (c_p)	$1004 \mathrm{\,J\,K^{-1}\,kg^{-1}}$
specific heat at constant volume (c_v)	$717 \mathrm{\,J\,K^{-1}\,kg^{-1}}$

Water substance

molecular weight (M_w)	$18.015 \mathrm{\,g\,mol^{-1}}$
gas constant for water vapor (R_w)	$461.5 \mathrm{\,J\,K^{-1}\,kg^{-1}}$
density of liquid water at $0\,°\mathrm{C}$	$1.000 \times 10^3 \mathrm{\,kg\,m^{-3}}$
latent heat of vaporization at $0\,°\mathrm{C}$	$2.500 \times 10^6 \mathrm{\,J\,kg^{-1}}$

STP	$T = 273.16 \mathrm{\,K}, p = 1013.25 \mathrm{\,hPa}$

Table A2 *Selected physical quantities and their units*

Quantity	Unit	Abbreviation
mass	kilogram	kg
length	meter	m
time	second	s
force	newton	N
pressure	pascal	$\mathrm{Pa} = \mathrm{N\,m^{-2}} = 0.01 \mathrm{\,hPa}$
energy	joule	J
temperature	degree Celsius	°C
temperature	degree Kelvin	K
speed		$\mathrm{m\,s^{-1}}$
density		$\mathrm{kg\,m^{-3}}$
specific heat		$\mathrm{J\,kg^{-1}\,K^{-1}}$

Table A3 *Selected conversions to SI units*

Quantity	Conversion
energy/heat	$4.186\,J = 1\,cal$
	$1\,kWh = 3.6 \times 10^6\,J$
pressure	$1\ atm = 760\,mm\,Hg$
	$1\,atm = 29.9213\,in\,Hg$
distance	$1\,m = 3.281\,ft$
temperature	$T(K) = T(°C) + 273.16$

Each side of an equation must have the same dimensions. This principle can often be used to find errors in a problem solution.

Appendix B

Notation and abbreviations

atm	pressure unit, one atmosphere (Chapter 1)
$\mathbf{a}, \mathbf{b}$, etc.	arbitrary vectors (Chapter 9)
$\mathbf{a}, a_x$	vector acceleration and its x component (Chapter 9)
a	Bohr radius (Chapter 2)
a	droplet radius (m) (Chapter 5)
a, b	empirical coefficients used in van der Waals equation of state (Chapter 5)
a^*	critical droplet radius (m) (Chapter 5)
a_x, a_y, a_z	the Cartesian components of vector $\mathbf{a}$ (Chapter 9)
$\mathbf{A}$	a vector field (Chapter 9)
A	horizontal area of a slab (m^2) (Chapter 6)
c	speed of light in vacuum (m s^{-1}) (Chapter 8)
c_v, c_p	specific heats (heat capacity per kg) at constant volume, pressure (J kg^{-1} K^{-1}) (Chapter 3)
$\bar{c}_v, \bar{c}_p$	molar specific heats (J mol^{-1} K^{-1}) (Chapter 3)
$\bar{c}_v, \bar{c}_p, \bar{G}$	overbar indicates quantities expressed per mole (Chapter 4)
CAPE	convective available potential energy (J kg^{-1}) (Chapter 7)
CIN	convective inhibition energy (J kg^{-1}) (Chapter 7)
C_v, C_p	heat capacities at constant volume, pressure (J K^{-1}) (Chapter 3)
div$_2\mathbf{V}$	divergence of vector field $\mathbf{V}$ in two dimensions (Chapter 9)
dH/dt	rate of heating (Chapter 9)
$dH/dt, dQ/dt$	time rate of change of enthalpy, heating rate (J s^{-1}) (Chapter 3)
$(dH)_p$	change in enthalpy at constant pressure (J) (Chapter 3)
dQ_{rev}	infinitesimal absorption of heat, subscript indicating that the change is reversible (J) (Chapter 4)
dQ_M/dt	rate of heating per unit mass (J kg^{-1} s^{-1}) (Chapter 9)
$đQ, đW$	differentials for heat, work, bar emphasizes path dependence (J) (Chapter 3)
$dS_{U,V}$	infinitesimal change in entropy during which U and V are held constant (J K^{-1}) (Chapter 4)

D/Dt	material derivative (Chapter 9)
$\Delta_f \overline{H}^\circ(X)$	standard heat of fusion of substance X ($J\,mol^{-1}$) (Chapter 3)
$\Delta \overline{G}^\circ$	change in Gibbs energy per mole, at standard conditions ($kJ\,mol^{-1}$) (Chapter 8)
ΔV	change in volume (m^3) (Chapter 3)
$\Delta_{vap} \overline{H}^\circ(X)$	standard heat of vaporization of substance X, overbar indicates 1 mol of substance, superscript o indicates at standard conditions (usually 25°C) ($J\,mol^{-1}$) (Chapter 3)
Δx	displacement in x (Chapter 3)
$\Delta \overline{H}^\circ$	change in enthalpy per mole, at standard conditions ($kJ\,mol^{-1}$) (Chapter 8)
Δt	time interval (s) (Chapter 2)
e, e_s	vapor pressure, saturation vapor pressure (Pa) (Chapter 5)
e'	vapor pressure over a solution (Pa) (Chapter 5)
E_{act}	activation energy (J) (Chapter 8)
E	energy (J) (Chapter 8)
ϵ	$\epsilon = M_w/M_d = 0.622$ (dimensionless) (Chapter 5)
η	mixing ratio (Chapter 8)
f	frequency of electromagnetic wave (Hz) (Chapter 8)
f	number of degrees of freedom of a molecule (Chapter 3)
f_x	partial derivative with respect to x (Chapter 3)
F	force (N) (Chapters 2, 3)
F	energy flux ($J\,s^{-1}$) (Chapter 8)
$F_\lambda(z)$	flux of electromagnetic energy at wavelength λ, elevation z ($J\,s^{-1}\,m^{-1}$) (Chapter 8)
$F, F/\mathcal{M}$	force, per unit mass ($N\,kg^{-1}$) (Chapter 7)
$\mathbf{F}$	force (N) (Chapter 4)
g, l, aq	gaseous, liquid, aqueous phase (Chapter 8)
g	acceleration due to gravity ($9.81\,m\,s^{-2}$) (Chapter 6)
g	Gibbs energy per kilogram ($J\,kg^{-1}$) (Chapter 4)
g, g_z, g_0	acceleration due to gravity, its value as a function of altitude, its value at the surface ($m\,s^{-2}$) (Chapter 1)
g_l, g_g	specific Gibbs energy for liquid, gas ($J\,kg^{-1}$) (Chapter 5)
g_v, g_w	specific Gibbs energy for vapor, liquid water ($J\,kg^{-1}$) (Chapter 5)
G	gravitational constant (Chapter 1, Table A1)
G	Gibbs energy (J) (Chapters 4, 8)
$\overline{G}$	Gibbs energy per mole ($J\,mol^{-1}$) (Chapter 4)
γ	ratio of specific heats c_p/c_v (dimensionless) (Chapter 3)

$\Gamma_d, \Gamma_m, \Gamma_e$	lapse rate, $-dT/dz$ of dry air ascending adiabatically, of moist adiabat, of the environment ($K\,m^{-1}$) (Chapter 6)
∇T	$\nabla T = (\partial T/\partial x)\mathbf{i} + \cdots$, gradient of T (Chapter 9)
h	height (Chapter 1)
h	height above a reference level (Chapter 6)
h	specific enthalpy ($J\,kg^{-1}$) (Chapter 5)
h	Planck's constant ($J\,s$) (Chapters 2, 8)
h, u	specific enthalpy, internal energy ($J\,kg^{-1}$) (Chapter 7)
h_0	initial height (Chapter 2)
H	enthalpy (J) (Chapters 3, 4)
H	scale height (Chapter 6)
$H(r)$	flux of heat crossing the surface of a sphere ($J\,s^{-1}$) (Chapter 5)
H_a	scale height of the atmosphere (Chapter 3)
H_w	a scale height for water vapor (m or km) (Chapter 6)
$\overline{H}^\circ$	standard enthalpy ($kJ\,mol^{-1}$) (Chapter 8)
$\mathbf{i, j, k}$	the Cartesian unit vectors (Chapter 9)
J	photodissociation coefficient (s^{-1}) (Chapter 8)
k	reaction coefficient (Chapter 8)
k_B	Boltzmann's constant (Chapters 1, 2, 3, Table A1)
K, K_p	equilibrium constant (Chapter 8)
K_H	constant in Henry's Law (Chapter 8)
$\mathcal{K}$	kinetic energy (J) (Chapter 7)
$\kappa = R/c_p$	(dimensionless) (Chapters 3, 4, 6) (0.286 for dry air)
κ_H	thermal conductivity coefficient ($J\,m^{-1}\,s^{-1}\,K^{-1}$) (Chapters 3, 5, 9)
LCL	lifting condensation level (Chapter 5)
$\mathbf{L}$	dimension length (m) (Chapter 1)
L	enthalpy of evaporation (latent heat) (Chapter 9)
L	length of box edge (m) (Chapter 2)
λ	mean free path (m) (Chapter 2)
λ	wavelength of an electromagnetic wave (m) (Chapter 8)
$L = \Delta H_{vap}$	the enthalpy (latent heat) of evaporation ($J\,kg^{-1}$) (Chapters 5, 6)
L_{evap}	latent heat of evaporation ($J\,kg^{-1}$) (Chapter 7)
mb	pressure unit, one millibar $= 1\,hPa$ (Chapter 1)
m_e	electron mass (kg) (Chapter 2)
m_0	mass of a molecule (kg) (Chapter 2)
$\mathbf{M}$	dimension mass (kg) (Chapter 1)
M_{eff}	effective molecular weight ($g\,mol^{-1}$) (Chapter 2)
M_G	gram molecular weight of a gas ($g\,mol^{-1}$) (Chapter 2)
M_v, M_d, M_e	gram molecular weight of vapor, dry air and effective ($g\,mol^{-1}$) (Chapter 5)

$\mathcal{M}$	mass (kg) (Chapters 2, 3, 4)
$\mathcal{M}$	mass of an object or system (Chapter 1)
$\mathcal{M}_E$	mass of the Earth (Chapter 1)
$\mathcal{M}_l, \mathcal{M}_g$	bulk mass of liquid, gas (kg) (Chapter 5)
n	unit vector (Chapter 9)
$\mathbf{n}_\theta$	unit vector in the theta direction (polar coordinates) (Chapter 9)
n_s	number density of vapor molecules at saturation (molecules m^{-3}) (Chapter 5)
n_{sat}	number density of vapor molecules at saturation (molecules m^{-3}) (Chapter 5)
n_w	number density of water molecules in vapor (molecules m^{-3}) (Chapter 5)
$n_0(z)$	molecular density as function of height (molecules m^{-3}) (Chapter 2)
n_0	number density of photons (photons/m^{-3}) (Chapter 8)
n_0	number density (molecules m^{-3}) (Chapters 3, 5)
N	newtons (Chapter 2)
N	total number of molecules (Chapter 2)
N_A	Avogadro's number (molecules mol^{-1}) (Chapters 1, 2, 5, Table A1)
$N(z)$	concentration of absorbers at level z (absorbers m^{-3}) (Chapter 8)
ν, ν_A, ν_B	number of moles, number of moles of species A, B (Chapters 3, 4)
ω, f	angular frequency (rad s^{-1}), frequency (cycles s$^{-1} \equiv$ Hz) (Chapter 6)
p	pressure (Pa) (all chapters)
p, p_G	pressure, partial pressure for gas G (Chapter 2)
$p, p(z), p_0$	pressure, as a function of z, at a reference level (hPa) (Chapter 6)
$p(V)$	pressure as a function of volume; expression for a curve in the p–V plane (Chapter 4)
Pa	unit of pressure, 1 Pa $=$ N m^{-2} (Chapter 1)
$\mathcal{P}(z)$	probability density function (Chapter 2)
$\mathcal{P}(v_x, v_y, v_z)$	joint probability density function for velocity components (Chapter 2)
$\Psi(z), \Psi_1, \Psi_2$	geopotential height as a function of height, at two levels (meters, on charts often in decameters, dm) (Chapter 6)
q	specific humidity (kg water vapor/kg of moist air) (Chapter 5)
r	$\mathbf{r} = x\mathbf{i} + y\mathbf{j} + z\mathbf{k}$ position vector (Chapter 9)
r	unit vector in the **r** direction (polar coordinates) (Chapter 9)
r	relative humidity (Chapter 5)

r_0	effective molecular radius (m) (Chapter 2)
R	gas constant for a particular gas ($J\,kg^{-1}\,K^{-1}$)
R_d	gas constant for dry air ($J\,kg^{-1}\,K^{-1}$) (Chapter 2, Table A1)
R_{eff}	effective gas constant for a mixture of species ($J\,kg^{-1}\,K^{-1}$) (Chapters 2, 5)
R_w	the gas constant for water vapor (Chapters 2, 5, Table A1)
R^*	universal gas constant ($J\,mol^{-1}\,K^{-1}$) (Table 1.1)
ρ	density ($kg\,m^{-3}$) (all chapters)
ρ, ρ_0, ρ_e	density, at a reference level, of the environment ($kg\,m^{-3}$) (Chapter 6)
ρ_a, ρ_e	density for adiabat and environment ($kg\,m^{-3}$) (Chapter 7)
s	entropy per unit mass (lower case indicates per unit mass) ($J\,K^{-1}\,kg^{-1}$) (Chapter 4)
s_l, s_g	specific entropy for liquid, gas ($J\,K^{-1}\,kg^{-1}$) (Chapter 5)
SI	Standard International system of units (Chapter 1)
$\mathbf{S}, d\mathbf{S}$	surface, surface element vector (Chapter 9)
S	entropy ($J\,K^{-1}$) (Chapter 5)
S, S_A, S_B	entropy, entropy of state A, state B ($J\,K^{-1}$) (Chapter 4)
$S_{sys}, S_{surr}, S_{universe}$	entropy for the system, surroundings, universe (sys+surr) (Chapter 4)
σ	standard deviation (Chapter 2)
σ	surface tension ($J\,m^{-2}$) (Chapter 5)
σ^2	variance (Chapter 2)
$\sigma(\lambda)$	absorption cross-section (m^2) (Chapter 8)
σ_c	collision cross-section (m^2, nm^2) (Chapter 2)
t_0, t_c	both stand for lifetime (s) (Chapter 8)
$t_{1/2}$	half life (s) (Chapter 8)
T	Kelvin temperature (K) (Chapters 2, 3, 5)
T_C	temperature measured in degrees Celsius (°C) (Chapter 5)
T_D	dew point temperature (K) (Chapter 5)
T_v	virtual temperature (K) (Chapter 5)
T_w	wet-bulb temperature (K) (Chapter 5)
$\mathbf{T}$	dimension time, also period of a repeating process, and temperature (Chapter 1)
Temp	temperature dimension (Chapter 1)
$T, T(z), T_0$	temperature, as a function of z, at a reference level (K) (Chapter 6)
$T_e(z), T_a(z)$	temperature of environment, of an adiabat (K) (Chapter 6)
$\overline{T}$	vertical average temperature in a layer of air (K) (Chapter 6)

θ	potential temperature (K) (all chapters)
θ_e	equivalent potential temperature (K) (Chapter 5)
θ_w	wet-bulb potential temperature (K) (Chapter 5)
θ_s	saturation equivalent potential temperature (K) (Chapter 5)
Θ	zenith angle (Chapter 8)
τ	optical depth (dimensionless) (Chapter 8)
(u, v, w)	the velocity components (v_x, v_y, v_z) (Chapter 2)
U	internal energy (J) (all chapters)
U_A, U_B	internal energy at states A, B (J) (Chapter 4)
$\mathbf{v}$	three-dimensional velocity vector field (m s^{-1}) (Chapter 9)
v	speed (m s^{-1}), sometimes with a subscript indicating velocity component along a coordinate axis (e.g., v_x) (Chapter 1)
v_{esc}	escape velocity (km s^{-1}) (Chapter 2)
v_l, v_g	specific volume for liquid, gas (m^3 kg^{-1}) (Chapter 5)
v^*	threshold velocity to exceed an activation barrier (m s^{-1}) (Chapter 8)
$\overline{v}$	mean speed (m s^{-1}) (Chapter 2)
$\overline{v}$	mean molecular speed (m s^{-1}) (Chapter 5)
$\overline{v^2}$	mean square velocity (m^2 s^{-2}) (Chapter 2)
$\overline{v_x^2}$	mean square of x component of velocity (m^2 s^{-2}) (Chapter 2)
$\mathbf{V}$	upper case usually indicates two-dimensional velocity (Chapter 9)
V	volume of a system (m^3) (Chapters 1, 3)
V	volume (m^3) (Chapter 3)
V_a, V_e	volumes of a parcel along an adiabat and of the environment (m^3) (Chapter 7)
V_A, V_B	initial and final volumes (Chapter 3)
w, w_s	mixing ratio, saturated (kg water vapor per kg dry air) (Chapters 5, 6, 7)
$W_{A \to B}, Q_{A \to B}$	work done by the system, heat taken into the system in going from state A to state B (J) (Chapter 4)
$W_{V_1 \to V_2}$	work in going from V_1 to V_2 (Chapter 3)
$\mathcal{W}, \mathcal{Q}$	work done by the system, heat taken into the system (Chapter 3)
$[X]$	concentration of X (Chapter 8)
$[X]_0$	concentration of X at $t = 0$ (Chapter 8)
z	elevation (m) (Chapter 2)
$z, \Delta z$	vertical distance, increment of it (m) (Chapter 6)
Z_p	altitude of pressure level p (m) (Chapter 6)

Appendix C

Answers for selected problems

Chapter 1

1.1 26 428 ft for $H = 8$ km.

1.2 0.0013.

1.3 819 hPa (using $H = 8$ km, 1 mile $=1.6$ km).

1.4 (a) 40 200 km, (b) 111.7 km deg^{-1}, (c) 96.7 km deg^{-1}.

1.5 50 000 km^2. 43 000 km^2.

1.8 11.2 km s^{-1}.

1.9 $2\sqrt{2h/g}$.

Chapter 2

2.1 1.29, 1.25, 1.2 kg m^{-3}.

2.2 0.80 kg m^{-3}.

2.3 212 hPa.

2.4 12.7 N.

2.5 2.69×10^{25} molecules m^{-3}.

2.6 $R_d = 2.87$ hPa K^{-1} m^3 kg^{-1}.

2.7 Differentiate $\mathcal{P}(v)$ with respect to v and set it to zero. See Table 2.1.

2.8 462, 493, 413, 1846 ms^{-1}.

2.9 630 Pa. 1 kg m^{-1}. Pressure with moisture $= 78646$ Pa. Density of moist mix $=$ 1 kg m^{-3}. Density of dry air at same temperature and pressure: 1.003 kg m^{-3}.

2.10 1.6×10^8 J.

2.11 4.3×10^7 J.

2.12 $n_0 = 2.69 \times 10^{25}$ molecules m^{-3}, $H = 8000$ m, $N = 2.15 \times 10^{29}$ molecules.

2.13 $z_{\lambda=H} = H \ln(n_0 \sigma_c H)$.

2.14 $(gp_0/RT_0) \int_0^\infty z e^{-z/H} dz = gp_0 H^2/RT_0$.

2.15 $\frac{1}{2}Nm_0\overline{v^2} = \frac{3}{2}Nk_BT$, see Problem 2.12 for N.

2.16 $f_{\text{yellow}} = c/\lambda = 6.0 \times 10^{14}$ s^{-1}. $f_{\text{coll}} = n_0 \sigma_c \overline{v} \approx 6.5 \times 10^9$ s^{-1}. An excited atom might suffer tens of collisions before it relaxes to the lower energy level.

Chapter 3

3.1 (d) $p_0 e^{-1/2}/2H$.

3.2 (a) ρR. (b) and (c) Let $1/T \equiv x$, then take the partial with respect to x.

3.3 $\kappa_T = 1/p$; to get κ_θ use $pV^\gamma = p_0 V_0^\gamma$; result $1/\gamma p$.

3.4 $\beta = 1/T$.

3.6 195.8 kJ.

3.7 $T = 205.1$ K. $\Delta T = -67.9$ K. Therefore, $\Delta U = 48.7$ J $= W$, since $Q = 0$.

3.8 (a) 10 kJ, (b) 7.17 kJ, (c) $dU + V\,dp \to dU$.

3.9 (a) 59.7 kJ, (b) 59.7 kJ.

3.10 $1.202\ \text{kg m}^{-3}$. $Q = -7500$ J. Use the mass of the dry air only as an approximation. $\Delta T = 8.7$ K.

3.11 $T_{600} = 248.7$ K. $\Delta U = M c_v \Delta T$, $\Delta H = -21.4$ kJ.

3.12 $0.02\ \text{K kg}^{-1}\text{s}^{-1}$, $20\ \text{J kg}^{-1}\text{s}^{-1}$, ΔU per unit time $= 14.3\ \text{J kg}^{-1}\text{s}^{-1}$.

3.13 (a) 12,500 kg. (b) Mass of an "air" molecule $= 0.029/N_A = 4.82 \times 10^{-26}$ kg. $N = 2.6 \times 10^{29}$ molecules. (c) 1.3×10^{44}.

3.14 $v_{\text{sound}} = \sqrt{RT}$ for isothermal; $\sqrt{\gamma RT}$, $\gamma = 1.40$ for adiabatic compression waves.

3.15 $\theta = T_0 (p_0/p)^{0.286}$.

Chapter 4

4.1 Zero if reversible.

4.2 $23.5\ \text{J K}^{-1}$.

4.3 $\Delta\theta = 65.8$ K.

4.4 $\Delta H = 500$ J. $\Delta T = 0.5$ K. $\Delta S = 2\ \text{J K}^{-1}$.

4.5 (a) $(\Delta S)_a = 23.5\ \text{J K}^{-1}$, (b) $(\Delta S)_b = 120.6\ \text{J K}^{-1}$, (c) $(\Delta S)_c = 8.6\ \text{J K}^{-1}$.

4.6 (b) $V_2 = 3.44\ \text{m}^3$, $T_2 = 227$ K, (c) $\Delta H = -146.7$ kJ, $\Delta U = -104.7$ kJ, (d) $W = 104.7$ kJ.

4.7 (a) $p_2 = 37877$ Pa, $W_2 = -65.2$ kJ, (b) $W_1 = 104.7$ kJ, $W_2 = -65.2$ kJ, $W_{\text{tot}} = 39.5$ kJ, (c) $T_3 = 113.6$ K, $(\Delta S)_2 = 1.39\ \text{kJ K}^{-1}$.

4.8 Use the isotherm joining the beginning and end points. $\Delta S = 199\ \text{J K}^{-1}$. $\Delta U = \Delta H = 0$.

4.9 Both zero for the cycle, if it is reversible. T_{max} is at the point A. T_{min} is at the opposite point on the circle.

4.10 Sliding up and down an isotherm in the $T - S$ diagram encloses no area.

4.14 -19.2 kJ.

Chapter 5

5.1 (a) 2.5×10^6 J, (b) $9.15\,\mathrm{J\,K^{-1}}$.

5.2 283.6 K.

5.3 $e_s = 65\,\mathrm{hPa}, z = 22\,\mathrm{km} \times \ln(1000/66.75) = 21.7\,\mathrm{km}$.

5.4 1.4%.

5.5 3.8 cm.

5.6 (a) 23.6 hPa, (b) 11.8 hPa, (c) $1.171\,\mathrm{kg\,m^{-1}}$, (d) $w = 0.0074\,\mathrm{kg\,kg^{-1}}$, $w_s = 0.0149\,\mathrm{kg\,kg^{-1}}$, (e) w.

5.7 (0.01, 6.80 hPa, 112.8%), (0.10, 6.10 hPa, 101.2%), (1.0, 6.03 hPa, 100.1%), (10, 6.025 hPa, 100.0%).

5.8 $e^{b/a} = 1.002$.

5.9 101.4%.

5.10 Radius at the maximum of the Köhler curve $a' = \sqrt{3B/b}$; RH $= (1 + 2b^{1.5}/3^{1.5}B^{0.5}) \times 100\%$; $a' = 6\,\mu\mathrm{m}$, RH $= 100\%$.

Chapter 6

6.1 $p(z) = p_0\exp\left(-(gz_0/RT_0)e^{(z/z_0-1)}\right)$.

6.2 $0.12\ \mathrm{m\ s^{-2}}$.

6.3 $0.6\ \mathrm{m\ s^{-1}}$, $1.2\ \mathrm{m\ s^{-1}}$, $3.6\ \mathrm{m\ s^{-1}}$, 3 min.

6.4 Adiabatically $0.03\ \mathrm{m\ s^{-2}}$, isothermally $0.02\ \mathrm{m\ s^{-2}}$.

6.5 If $p_0 = 1000$ hPa, $T_0 = 300$ K, $\Gamma = 6$ K km^{-1}, $z_{10} = 10$ km, $T_1 = T_0 - \Gamma z_{10} = 240$ K, then

$$
p(z) = \begin{cases}
p_0 \left(\frac{T_0 - \Gamma z}{T_0}\right)^{g/R\Gamma}, & z < z_{10} \\[2mm]
p_0 \left(\frac{T_1}{T_0}\right)^{g/R\Gamma} e^{-g(z-z_{10})/RT_1}, & z > z_{10}.
\end{cases}
$$

6.6 $\Theta = T_0 \exp\left(-z/z_0 + \kappa g z_0(e^{z/z_0-1})/RT_0\right)$.

6.7 $\Gamma \approx 5$ K km^{-1}.

6.8 $Z(p) = -\frac{Ra}{g} \ln \frac{p}{p_0} - \frac{Rb}{2gp_0}(\ln \frac{p}{p_0})^2$.

6.9 926 m.

6.10 $0.516\ \mathrm{m\ s^{-2}}$.

6.11 $\omega = 5.7 \times 10^{-3}$ rad s^{-1}, $f = 9.1 \times 10^{-4}$ Hz, $T = 18$ min.

6.13 (a) $A = 10$ m, $B = 0$, (b) $A = 0$, $B = 175$ m.

Chapter 7

7.1 (a) $16\,\mathrm{g\,kg^{-1}}$, (b) $22\,\mathrm{g\,kg^{-1}}$, RH $= 73\%$, (c) 20 °C, (d) 940 hPa, (e) 23 °C, (f) yes, (g) yes, (h) $8.5\,\mathrm{g\,kg^{-1}}$, (i) $12\,\mathrm{g\,kg^{-1}}$.

7.2 (a) 270 hPa, (b) cold and dry, (c) very low to 400 hPa, above that $T - T_D \approx$ 10 °C, (d) between 950 and 910 hPa, 750 and 700 hPa, above 200 hPa, (e) no, yes.

7.3 (a) 150 hPa, (b) unstable, (c) moderate below 700 hPa, dry above, (d) at 650 hPa, (e) large CAPE, no CIN, unstable.

7.4 (a) $w = 3.6 \text{g kg}^{-1}$, RH $= 45\%$, $\theta = 284$ K, (b) 840 hPa, (c) $\theta_e = 294$ K, (d) the same as in (a), (e) the same as in (c).

7.5 (a) 3.5 g kg^{-1}, (b) $T = 30 °C$, $T_D = -30 °C$. Note: the numbers here are unrealistic but easy to work with on the charts.

Chapter 8

8.1 $-33.81 \text{ kJ mol}^{-1}$, $-391.9 \text{ kJ mol}^{-1}$; both exothermic.

8.2 $\overline{H}^\circ = 72.14 \text{ kJ mol}^{-1}$, no.

8.3 Low temperatures for forward, high for reverse (this is how NO is formed in combustion).

8.4 $3.4 \times 10^{-19} \text{ J} = 2.1 \text{ eV}$.

8.5 6×10^{34}, Lower the temperature.

8.6 k_1/k_2.

8.7 $2.68 \times 10^{-4} \text{ mol l}^{-1}$.

8.8 2.5×10^{-2} mol.

Chapter 9

9.1 $d(a_x b_x)/dt + d(a_y b_y)/dt = b_x \, da_x/dt + a_x \, db_x/dt + \cdots$.

9.2 $|\mathbf{v}| \cos 30°\mathbf{i} + |\mathbf{v}| \sin 30°\mathbf{j} = |\mathbf{v}|\frac{\sqrt{3}}{2}\mathbf{i} + |\mathbf{v}|\frac{1}{2}\mathbf{j}$.

9.3 $\frac{\sqrt{2}}{2}(\mathbf{i} + \mathbf{j})$, $\frac{\sqrt{2}}{2}(\mathbf{i} - \mathbf{j})$.

9.4 $-0.239\mathbf{i} + 0.160\mathbf{j} - 0.958\mathbf{k}$.

9.5 Circle with radius α centered at (x_0, y_0). Point on the rim of a rolling wheel.

$$\mathbf{r} = (x_0 + \alpha \cos \omega t + \gamma t)\mathbf{i} + (y_0 + \alpha \sin \omega t)\mathbf{j}$$

$$\mathbf{v} = (\gamma - \omega\alpha \sin \omega t)\mathbf{i} + \omega\alpha \cos \omega t\mathbf{j}.$$

9.6 Three maxima.

$$dp = \frac{3}{R}A \cos \frac{3x}{R} \cos \frac{\pi y}{2L} e^{-z/H} \, dx - \frac{\pi}{2L}A \sin \frac{3x}{R} \sin \frac{\pi y}{2L} e^{-z/H} \, dy$$
$$- \frac{A}{H} \sin \frac{3x}{R} \cos \frac{\pi y}{2L} e^{-z/H} \, dz.$$

To get the gradient replace (dx, dy, dz) by $(\mathbf{i}, \mathbf{j}, \mathbf{k})$.

9.7 $u = -yv_0/r^{\ell+1}, v = xv_0/r^{\ell} + 1.$ div$_2(\mathbf{v}) = 0$, since $v_r = 0, v_\theta$ has no θ dependence.

9.8 Elliptically shaped bowl. $\nabla p = 4a(x - x_0)^3\mathbf{i} + 2b(y - y_0)\mathbf{j}, \partial p/\partial n = \frac{\sqrt{2}}{2}(4a(x - x_0)^3 + 2b(y - y_0)).$

9.9 This is a wave that travels west to east and takes one month to circle the globe.

9.10 $0.01\,\mathrm{K\,s^{-1}}, \partial T/\partial t = 0.009\,\mathrm{K\,s^{-1}}.$

9.11 Warming: $0.00020\,\mathrm{K\,m^{-1}}$, same at fixed altitude, since $dT/dz = 0$.

9.12 (b) $1.1 \times 10^{-5}\,\mathrm{K\,s^{-1}}$, (c) $-1.3 \times 10^{-5}\,\mathrm{K\,s^{-1}}.$

Bibliography

Atkins, P., 1994. *Physical Chemistry*, New York: W. H. Freeman.

Atkins, P. and J. de Paula, 2002. *Atkins' Physical Chemistry*, seventh edition, Oxford: Oxford university Press.

Bohren, C. F. and B. A. Albrecht, 1998. *Atmospheric Thermodynamics*, New York: Oxford University Press.

Brasseur, G. and S. Solomon, 2005. *Aeronomy of the Middle Atmosphere*, Dordrecht: D. Reidel.

Brimblecombe, P., 1986. *Air Composition and Chemistry*, Cambridge: Cambridge University Press.

Callen, H. B., 1985. *Thermodynamics and an Introduction to Thermostatistics*, New York: Wiley.

Çengal, Y. A. and M. A. Boles, 2002. *Thermodynamics: An Engineering Approach*, New York: McGraw-Hill.

Curry, J. A. and P. J. Webster, 1999. *Thermodynamics of Atmospheres & Oceans*, San Diego, CA: Academic Press.

Denbigh, K., 1981. *The Principles of Chemical Equilibrium*, New York: Cambridge University Press.

Emanuel, K. A., 1994. *Atmospheric Convection*, New York: Oxford University Press.

Emanuel, K. A., 2005. *Devine Wind*, New York: Oxford University Press.

Fermi, E., 1956. *Thermodynamics*, New York: Dover.

Finlayson-Pitts, B. J. and J. N. Pitts, 2000. *Chemistry of the Upper and Lower Atmosphere*, San Diego, CA: Academic Press.

Fleagle, R. G. and J. A. Businger, 1980. *An Introduction to Atmospheric Physics*, San Diego, CA: Academic Press.

Giancoli, D. C., 2004. *Physics: Principles with Applications,* sixth edition, Englewood Cliffs, NJ: Prentice Hall.

Guggenheim, E. A., 1959. *Thermodynamics: An Advanced Treatment for Chemists and Physicists*, Amsterdam: North-Holland.

Halliday, D., R. Resnick and J. Walker, 2004. *Fundamentals of Physics,* seventh edition, New York: Wiley.

Hobbs, P., 2000. *Basic Physical Chemistry for the Atmospheric Sciences*, second edition, Cambridge: Cambridge University Press.

Holton, J. R., 1992. *An Introduction to Dynamical Meteorology*, New York: Academic Press.

Houston, P. L., 2001. *Chemical Kinetics and Reaction Dynamics*, New York: McGraw-Hill.

Houze, R. A., Jr., 1993. *Cloud Dynamics*, New York: Academic Press.

Irebarne, J. V. and W. L. Godson, 1981. *Atmospheric Thermodynamics*, Boston, MA: Kluwer.

Present, R. D., 1958. *Kinetic Theory of Gases*, New York: McGraw-Hill.

Reiss, H., 1965. *Methods of Thermodynamics*, Mineola, NY: Dover.

Rogers, R. R. and M. K. Yau, 1989. *A Short Course in Cloud Physics*, Oxford: Pergamon.

Sears, F. W., 1953. *An Introduction to Thermodynamics, the Kinetic Theory of Gases and Statistical Mechanics*, Reading, MA: Addison-Wesley.

Sears, F. W. and M. W. Zemanski, 1984. *University Physics*, sixth edition, Reading, MA: Addison-Wesley.

Seinfeld, J. S. and S. N. Pandis, 1998. *Atmospheric Chemistry and Physics*, New York: Wiley Interscience.

Stewart, J., 1995. *Calculus*, third edition, Pacific Grove, CA: Brooks/Cole.

Stull, R. B., 1988. *An Introduction to Boundary Layer Meteorology*, Dordrecht: Kluwer.

Stull, R. B., 2000. *Meteorology for Scientists and Engineers*, Pacific Grove, CA: Brooks/Cole.

Taylor, F. W., 2005. *Elementary Climate Physics*, Oxford: Oxford University Press.

US Air Weather Service Manual (AWSM 105-124).

Vardavas, I. M. and F. W. Taylor, 2007. *Radiation and Climate*, Oxford: Oxford University Press.

Wallace, J. M. and P. V. Hobbs, 2006. *Atmospheric Science: An Introductory Survey*, second edition, Amsterdam: Elsevier.

Warneck, P., 1999. *Chemistry of the Natural Atmosphere*, second edition, International Geophysics Series, San Diego, CA: Academic Press.

Whitten, K. W., R. E. Davis and L. Peck, 1996. *General Chemistry*, Orlando, FL: Saunders College Publishing.

Wills, A. P., 1958. *Vector Analysis with an Introduction to Tensor Analysis*, New York: Dover.

Zemansky, M. W., 1968. *Heat and Thermodynamics*, New York: McGraw-Hill.

Index